An Introduction to

WASTES MANAGEMENT

The Institution of Water and Environmental Management

ISBN 1 870752 01 5

The Institution of Water and Environmental Management
15 John Street, London WC1N 2EB

DEDICATION

By resolution of the Council of the
Institution of Water and Environmental Management
this Booklet on *Wastes Management*
is dedicated to the memory of the late

Bernard John Dangerfield

Editor to the Institution from 1987 to 1989 and for one of its predecessor bodies, the Institution of Water Engineers and Scientists, from 1969 to 1987.

Throughout his working life Mr Dangerfield was concerned with the editing and publishing of journals and other publications and his name is particularly associated with a series of water practice manuals for the Institution of Water Engineers and Scientists. Although he undertook with great success many other important responsibilities, publications were his first and abiding interest and he played a leading role in planning a new series of introductory booklets for IWEM of which this is the first.

FOREWORD

The Institution of Water and Environmental Management (IWEM) is a corporate learned society and examining body representing the interests of engineers and scientists and other professionally-qualified personnel engaged in the various sectors of the environment: water, air and land. It was formed in July 1987 by the unification of three eminent organizations, The Institution of Public Health Engineers, The Institution of Water Engineers and Scientists and the Institute of Water Pollution Control, each having a history of some 100 years.

Over the years the predecessor bodies have produced definitive manuals and other publications, notably in respect of British practice in the water industry. These have become reference sources for those who are actively engaged in the field, as well as for students seeking authoritative guidance in preparing for professional qualifications. Such publications are being continued by IWEM, and the range is being extended to take account of the wider environmental issues and interests which the new organization now embraces.

This *Introductory Booklet on Wastes Management* is the first of a series to be published by IWEM on an aspect of environmental management which is today giving rise to increasing concern, particularly in respect of the treatment and disposal of special and hazardous wastes. Although it is intended as an introduction to the subject, and has been written as a general guide to the interested layperson, the booklet provides a comprehensive summary of the situation in the UK, and should also be useful to those embarking on a career in the field.

The Institution wishes to record its thanks to those members who have contributed to the production of the booklet, and in particular to Mrs Jane Barron who has been responsible for the preparation of the text.

G. A. TRUESDALE
President

ACKNOWLEDGEMENTS

The Institution gratefully acknowledges permission from the following to reproduce plates:

Aspinwall & Company Ltd (Plate 6)

Cleanaway Limited (Plates 3, 5 and 7)

Leigh Environmental Ltd (Plate 8)

Surveyor Magazine (Plate 4)

T.T.B. (Fabrications) Limited (Plate 2)

Wimpey Waste Management Limited (Plate 1)

CONTENTS

ILLUSTRATIONS

PLATES

(*Between pages 30 and 31*)

TABLES

1. THE NEED TO MANAGE WASTES

Waste is something for which we have no further use and which we wish to get rid of. It can be solid or liquid, and includes waste products arising from our way of life and waste body products. Wastes therefore range from the materials which are discarded in household dustbins (tin cans, bottles, vegetable trimmings)—and flushed down our toilets (sullage, sewage)—to the by-products of chemical processes (brewing, pharmaceuticals, electroplating)—to agricultural waste (straw, manure)—and the wastes produced by the nuclear industry.

For the various categories of waste which are considered in this booklet, the total estimated arisings in the UK alone amount to over 480 million tonnes per year. By way of comparison this is almost equal to the total world production of iron, and one and a quarter times the world production of wheat. The sheer volumes of waste which are produced dictate that they cannot be discarded haphazardly—as may have been the case in the past; however, as with any other commercial industry which is responsible for the output of such large volumes, they need to be managed efficiently. In addition, the potentially-harmful nature of part of that product volume calls for control on a statutory basis.

Waste has previously been out of sight and therefore out of mind. However, public awareness of the danger to public health and the environment has led to (a) a requirement for knowledge about current practice, and (b) moves to control waste management in order to minimize the dangers. This trend commenced in the mid-nineteenth century and has continued, at an accelerating pace, ever since.

Waste is classified according to its source and according to the hazards associated with its handling and final disposal, i.e. danger to public health and the environment. Definitions and classifications of various wastes are given in the *Glossary.* Special classifications are given to wastes which present a particular hazard with respect to public health, i.e. hazardous, special and radioactive, and there are particular controls on the management of these wastes.

Wastes arising from different sources are variable in nature and require different treatment and disposal techniques. Fig. 1 shows that, whilst the vast majority of UK wastes are of agricultural origin, large quantities originate from mining and quarrying activities. Wastes arising from these sources are generally disposed of back to the land by the producer, and are not covered in any great detail in this booklet. Of the remaining wastes shown in Fig. 1 (just over 100 million tonnes), sewage sludge arisings are approximately equal to those of municipal waste (at about 25 million tonnes), and industrial wastes are about 45 million tonnes. Hazardous and special wastes scarcely

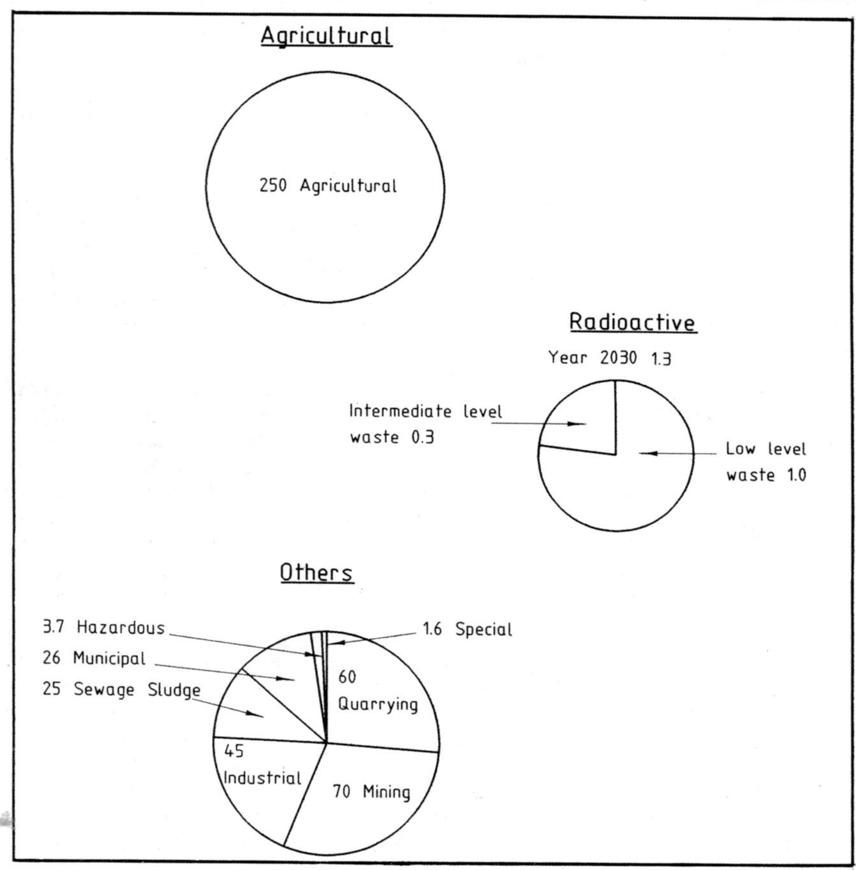

Fig. 1 Waste arisings (in million tonnes) during 1987

seem significant in terms of tonnage. It should also be noted from Fig. 1 that radioactive waste arisings, which will be stored for disposal until 2030, are even less than the special waste arisings. However, these small tonnages of waste (hazardous, special and radioactive) actually require more expertise for their handling and treatment and more stringent controls on their ultimate disposal, making them the most costly to manage (in terms of £/tonne).

The majority of wastes in the UK are disposed of to land, the sheer volume being useful for filling disused quarries and reclaiming marginal lands. However, although this is the preferred disposal route in this country, dangers to both public health and the environment can arise (see Section 4).

Public awareness of these dangers has been aroused by a number of incidents, the most notable of which in the UK have been the discovery by children of cyanide waste in the Midlands (in the early 1970s) and the destruction of part of a bungalow in Derbyshire (in 1986) which was caused by an explosion arising from the presence of landfill gas. Many people also know of the story of Love Canal in the USA where a residential development was built over a toxic waste dump. Chemical contamination and toxic gases came to the surface, poisoning the ground, and were thought to give rise to miscarriages in pregnant women, and disease and medical defects in both children and adults, resulting finally in an evacuation of the area. Although such a scenario is unlikely in the UK, it has helped to increase public awareness of the hazards which are associated with waste disposal.

Certain recent incidents have caught the media's attention and have become emotive issues. In the early 1980s a hazardous waste incinerator at Bonnybridge in Scotland was blamed for discharging contaminants into the atmosphere, and for causing a high incidence of birth defects in cattle on a nearby farm. Also, emissions from the Sellafield nuclear plant have been blamed for a high incidence of leukaemia in children in the area. Whilst there can be no doubt that certain discharges from some industrial plants can contain harmful substances, such diseases are known statistically to cluster at random and, whilst emotion may readily link postulated cause with observed effect, in practice it is much harder to establish a true cause/effect relationship on the basis of currently-recognized facts.

In 1988 it was reported that ships carrying (in one case) domestic waste, and (in another) toxic waste, were cruising the world's oceans seeking acceptable disposal sites (on land) for their cargo. The UK has been seen by some as a suitable recipient country for USA-generated domestic waste, and various plans have been put forward to landfill* these wastes, generally linked to the production of gas as a commercial by-product. Whilst the Government has now intervened to prevent this happening, it remains keen to capitalize on the UK's technical expertise and physical resources, and to promote the importation of hazardous waste for treatment on commercial sites.

* *'Landfill' is the term used for the controlled disposal of wastes to land.*

A public awareness of the emotive issues which are associated with waste management serves to reinforce the concept that waste must be managed and controlled as

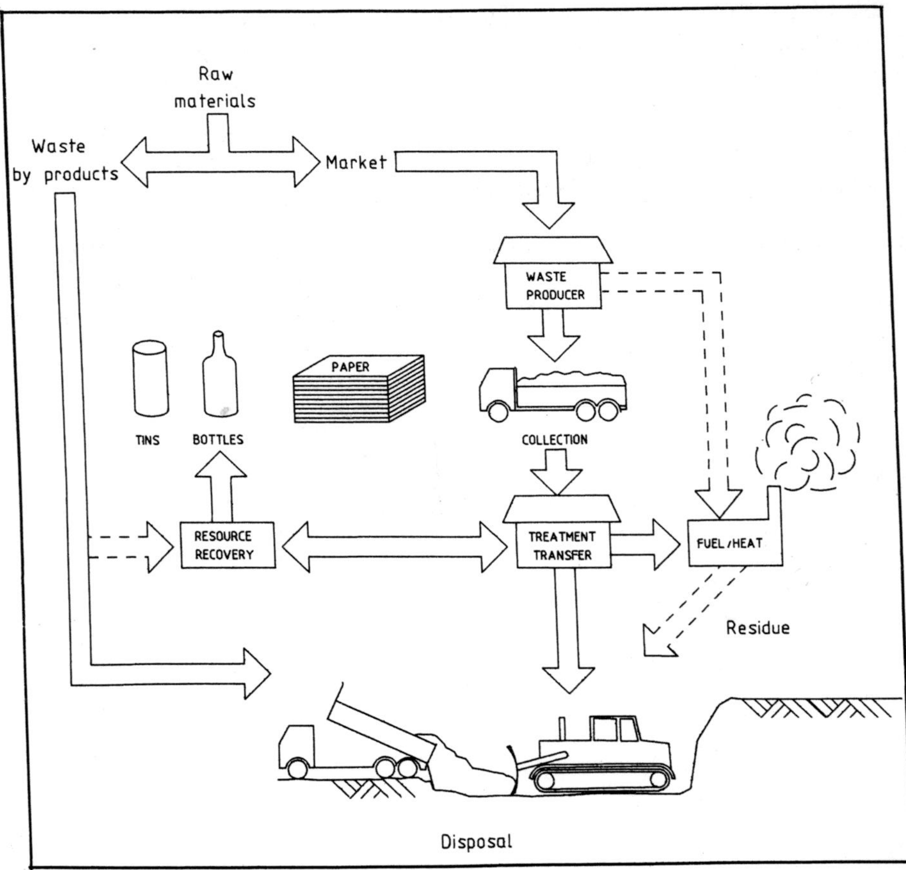

Fig. 2 Solid waste management: main stream process

rigorously as any commercial industry. Fig. 2 depicts the simplified process stream for the solid waste industry, from the incorporation of raw materials into a product to final disposal of the used article. Each step of the process must be carefully controlled to maximize efficiency and to comply with legislation which is intended to protect public health and the environment. However, attitudes towards protective legislation are influenced by the following factors:

(i) The actual, and the perceived, effectiveness of existing legislation;

(ii) Evidence of environmental harm both arising and potentially arising;

(iii) Public opinion and expressed environmental concern within the UK; and

(iv) Contemporary thinking and lobbying within the EC.

It was felt that it would be premature to anticipate the precise nature of proposed new legislation in this booklet. However, the Institution has a specialist group, i.e. the Environmental Group, which was formed to give advice on certain matters, specifically waste.*

It should be noted that whilst waste management practice in England and Wales is similar, it is different in Scotland. This is particularly true of the method of working of the waste disposal authorities and legislation.

* *Readers who are interested in up-to-date information should contact the Institution's Headquarters.*

2. ARISINGS, PLANNING AND LICENSING

INTRODUCTION

The magnitude of the problem associated with the management of waste materials only really becomes apparent on examination of the actual volumes or tonnages of these materials which are produced each year. When it is realized that, per head of population, about 1.5 tonnes of waste are discarded every year (i.e. the weight equivalent of a Ford Granada) and at an average density of about ⅓ tonne/cubic metre (about the same volume as two of these cars), and that there are about 55 million people in the UK, the size of the task can begin to be imagined. These figures only refer to typical household arisings, i.e. not to hazardous, radioactive, mining or quarry wastes, sewage sludge, or agricultural wastes.

ARISINGS

Of the categories considered in this publication (excluding agricultural and radioactive wastes) there were over 200 million tonnes of waste generated in 1987 in the UK (Fig. 1). Of this, about 26 million tonnes were municipal (household) arisings, about 4 million tonnes defined as hazardous and the remainder was mine and quarry, non-hazardous industrial waste and sewage sludge. Of the hazardous wastes about half were classified as liquids, but were not disposed of to sewer.

Controlled Wastes

In 1987 the responsibilities of waste disposal authorities (WDAs) were extended from arranging collection and disposal facilities for domestic, commercial and some industrial wastes to all controlled wastes as defined by the Control of Pollution Act, 1974. Therefore in England waste disposal authorities must find a disposal route for about 24.4 million tonnes of waste, in Scotland 4.5 million tonnes, and in Wales 1.6 million tonnes.

Hazardous Wastes

At present there are about 3.7 million tonnes of hazardous wastes in England, 200 000 tonnes in Wales and 80 000 tonnes in Scotland. These figures include special waste arisings.

Special Wastes

More detailed information is available for special wastes than for hazardous wastes and the *Hazardous Waste Inspectorate's (HWI) Third Report,* published in June 1988, gives detailed figures. The total arising in England in 1986–87 amounted to 1 477 074 tonnes.

Counties in England with more than 100 000 tonnes of special wastes arising each year include Cleveland, Hertfordshire and Lancashire. However, these are not necessarily the counties where most special wastes are disposed of. Counties where more than 100 000 tonnes of special waste are disposed of each year include Cheshire, Cleveland, Durham, Essex, and the former metropolitan counties of Greater Manchester, West Midlands and West Yorkshire.

In Wales the total arising in 1986–87 was 82 698 tonnes, of which only the Lliw Valley District generated more than 7000 tonnes (21 382 tonnes in this District).

Figures for Scotland are somewhat limited, but the HWI made an estimate that 30 000 tonnes of special waste were generated in 1986.

A total of approximately 1.6 million tonnes of special waste were therefore generated in England, Scotland and Wales in 1986.

Industrial Wastes

This category includes waste from a wide variety of sources, but is estimated to include at least 23 million tonnes of general industrial waste, 3 million tonnes of building waste and about 14 million tonnes of power station waste.

Mine and Quarry Wastes

About 70 million tonnes of mining wastes and 60 million tonnes of quarry wastes are produced each year.

Sewage Sludge

Some 25 million tonnes of sewage sludge are produced annually at sewage-treatment works in the UK. The water content varies between 95 and 98 per cent, and the overall dry solids content is estimated at 1.25 million tonnes.

Agricultural Wastes

About 250 million tonnes of agricultural wastes are produced each year, which includes manures and straw arising from animal bedding.

Radioactive Wastes

Statistics of radioactive waste arisings are compiled by NIREX (see Section 3) and published in a summary report entitled *Radioactive Waste Arisings in the UK.* This provides an estimate of all low and intermediate-level wastes to the year 2030 and some decommissioning wastes beyond 2030, and includes all existing stocks and projected arisings. Estimated volumes for disposal to 2030 (at as yet unidentified sites) are 300 000 m^3 of intermediate and 1 000 000 m^3 of low-level wastes. Of these volumes 80 000 m^3 of the intermediate and 250 000 m^3 of the low-level wastes will result from decommissioning. A small additional amount is currently disposed of on site at the British Nuclear Fuel sites at Dounreay and Sellafield.

IMPORTS AND EXPORTS

All waste disposal authorities (WDAs) import and export waste to facilitate the best and most economic method of disposal. In general this practice is confined to movements within the UK mainland. It is self-evident that waste, particularly household and commercial, which is generated in the large metropolitan areas must, to a large extent, be disposed of outside those areas. Also, from the pattern of special waste arisings and disposals, it is evident that special wastes are exported to the county with the best facilities for their treatment or disposal. The HWI report gives details of special waste imports and exports for each county.

In addition to movement of waste within the UK, there is a considerable amount of waste which is exported from and imported into the country. Exports of waste from the UK are mainly for disposal or incineration at sea, and this movement is mainly restricted to sewage sludge and hazardous wastes, for which this is currently considered to be the best practicable environmental option.

So far, wastes imported to the UK have been restricted to hazardous waste for which there are good treatment and disposal facilities. For certain wastes this practice should be continued, and indeed encouraged, as the imported quantities are necessary to make hazardous waste incinerators viable, and they generate income for UK companies. However, because the environmental laws governing the disposal of certain wastes to landfill are more relaxed in the UK than in certain of our European neighbours, most of the imported waste currently goes, untreated, to landfill. This practice does not help to support the country's hazardous-waste facilities, uses up valuable landfill space and should be discouraged. Recently there have been moves by private enterprises to import domestic wastes (principally from USA) for landfill in the UK. This proposal caused such a public outcry that the Government was required to take steps to prevent it being carried out.

WASTE DISPOSAL PLANS

Under the Control of Pollution Act 1974 all WDAs are required to prepare a plan showing how they intend to deal with waste arisings and disposals in their area. Thus for household, commercial and industrial (including toxic) wastes each WDA must give information in the plan of volumes arising, public and private treatment, and disposal facilities and costs to the authority of waste management. The waste disposal plan should be reviewed regularly and updated as necessary.

Of a potential 198 plans for the UK, 117 were complete, 16 were in draft form referred to the Secretary of State and 65 were incomplete by March 1987, ten years after the date when they were first formally required. (In 1989, 23 of 79 WDAs in England had deposited their plans with the Secretary of State.) Of the ones that are complete the quality of information given in them is extremely variable: some give detailed statistics of waste arisings by type and area, together with detailed routes for treatment and disposal, both of the WDA's wastes and of wastes imported into the WDA's area. Others give only a global view, with no breakdown of statistics, routes or methods.

PLANNING APPLICATIONS

Under the Town and Country Planning Acts (see Section 6) it has been a requirement since 1941 that any waste facility must have planning permission. This includes transfer stations, treatment plants, recycling plants, civic-amenity sites and disposal sites. In this booklet incinerators are considered to be treatment plants. Refuse-derived fuel plants (production and users) also require planning permission.

A planning application will be considered by the local planning authority in whose area the proposed site lies. In England this is a county authority, in Wales a district authority, and in Scotland the district, regional or island authority. In England the application may have to be submitted to the district or borough planning office for the site area, and will then be passed to the county planning authority for consideration. In the former metropolitan areas the district authority (in London the borough council) considers the application. In Scotland all planning applications are dealt with by the district council, except in Highland, Dumfries and Galloway and the Borders Regional Councils. In these areas applications are considered by the region as there is no district planning function.

The application, which should be made on a form supplied by the planning authority, should be accompanied by a statement and drawings identifying the site and outlining the proposed operation.

There is a significant variation in the amount of detail required by the planning authorities across the country. Although some will only require a simple form stating the type of application with the barest information relating to the operation, others will require full details including the type and number of vehicles to be used on site (e.g. for a landfill operation). It is in the interests of the applicant that all relevant planning matters, i.e. those related to land use, are addressed. Aspects such as the adequacy of proposals to deal with leachate, landfill gas, the phasing of operations and restoration (in the case of a landfill) or architecture and landscaping (in the case of a structure) can be important factors for consideration by the local authority. Since July 1988 it has also been a legal requirement that an environmental statement accompanies certain planning applications pertaining to waste disposal.

Consultations with interested parties are made by the planning authority whilst considering the application. Interested parties include water authorities (or river authorities and water companies), Nature Conservancy Council, Council for the Protection of Rural England, Heritage Trust, and others as appropriate. If valuable mineral deposits lie beneath the site, the owner of the land or the rights of the deposit will also be consulted.

At the time the application is made, notice of the application must be posted on the site and in a local paper. A copy of the application must be lodged at a place where locals can inspect it, i.e. in a local library or school.

SITE LICENCES

Once planning approval has been given, a site licence should be applied for, where necessary. This applies to plant and equipment for treating and handling waste, including disposal to land. The licence applies to the actual operation and is issued by the WDA in the case of landfill or Ministry of Agriculture, Fisheries and Food (MAFF) and Department of Agriculture and Fisheries (DAFS) for Scotland, in the case of the land disposal of sewage sludge. The WDA licence sets conditions for the operation of the facility and will include stipulations on the type of waste for which it may be used, the operating hours and methods to be employed to minimize nuisance to site neighbours and to protect public health and the environment.

LICENCES FOR DISPOSAL AT SEA

The disposal of waste at sea also requires a licence which is issued by the MAFF/DAFS under the Food and the Environment Act 1985. With the North Sea Treaty requiring a ban on dumping at sea from 1992, it is now difficult to obtain a licence for a new proposal. However, as Britain is not intending to comply with the ban but merely restricting itself to existing practice, current operations are looked upon more favourably. Even so, MAFF have recently been restricting some of their licences, particularly in the length of term for which they are valid.

3. TREATMENT AND TRANSPORT

INTRODUCTION

The principal reasons for treating waste are to make it easier to transport and dispose of, and to facilitate resource recovery. Therefore for most wastes (domestic, commercial and some industrial) the main aims are (a) to reduce bulk, (b) to reduce the costs of transportation, and (c) to prolong the life of disposal facilities. For hazardous and special wastes the objective is to reduce their hazardous nature, both with regard to human contact in the short term and to protect the environment in the long term.

Environmental and economic pressures (mainly the former) have led to increased efforts to promote the recovery and re-use of as much of the resources as possible. This includes recovery of materials (tin, glass, etc) segregated by the consumer/producer from the bulk of waste prior to disposal, recovery of by-products from industrial processes, and minimizing the use of fossil fuels through heat exchange, district heating and refuse-derived fuel. Not all these efforts have been successful, as may be seen in the later sections of this chapter.

Proportions of domestic waste treated for volume reduction or resource recovery are shown in Table I.

Table I. Methods of treatment and disposal of municipal wastes (1988–89)

	Per cent
Landfill disposal without treatment	79
Incineration	6
Compaction/shredding/baling	13
Reclamation and other methods	2

TREATMENT OF BULK SOLID WASTES

Volume Reduction

The two main treatment methods which are available to reduce volume are shredding/pulverization and baling. Volume reduction by shredding/pulverization may be achieved by either a dry or wet process. The dry process is known as shredding and is carried out by a flail or hammer mill with a vertical shaft. This smashes the waste, reducing it in size, after which the components can be separated ballistically or by air classification. The wet process is known as pulverization, and all such processes in this country are based on the *Dano* process. This is a horizontal rotating drum into which waste and a lubricant (usually water) are fed at one end and waste, separated by size (usually a 34–45 mm diameter screen), emerges from the other end. The smaller material (known as the fines or the product) is generally organic in nature, and the larger material (known as the rejects or contraries) is generally inorganic (glass, tin, etc). A typical flow diagram for a pulverization process is shown in Fig. 3.

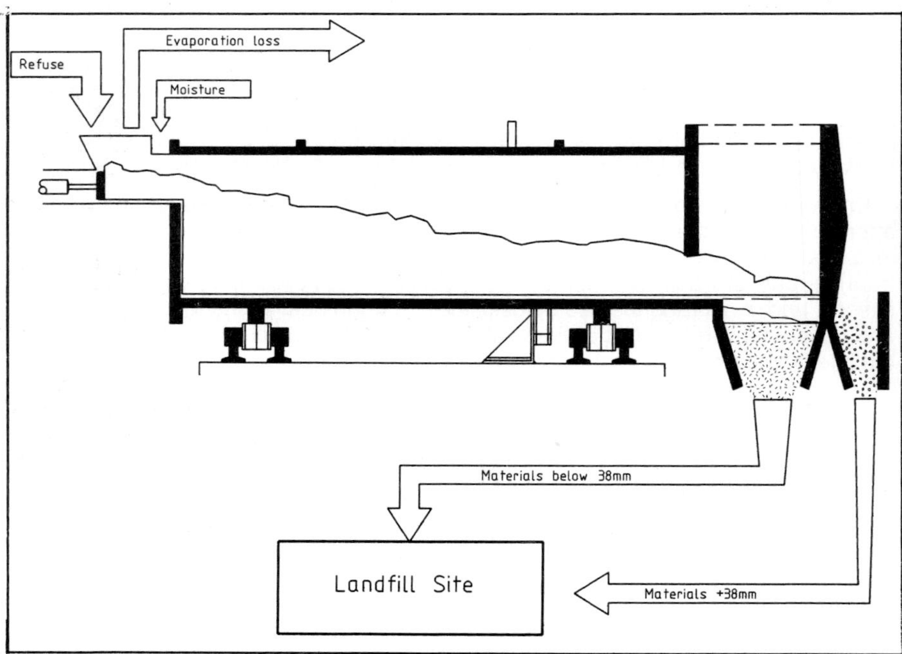

Fig. 3 Flow diagram of wet pulverization process

These two processes are popular (in principle) as a treatment method, since not only do they reduce the bulk volume of wastes for transportation but they separate out the organic and inorganic fractions. The organic part of the waste can then be made into compost or refuse-derived fuel, and the inorganic constituents can be further separated out (or classified) for resource recovery. Items such as paper and rag cannot be recovered in this way, and are generally hand-picked prior to the treatment process. However, although these processes may be popular in principle, practically they have been used with limited success in the UK, mainly because of the difficulty of finding an end-use for the treated or recovered materials.

Another method of treating wastes to reduce volume is by baling. This is extensively used in Scotland, and involves the use of three hydraulic rams compressing the waste into bales which are about 1 tonne in weight and 1 m^3 in volume. High-density bales are self-sustaining. A certain amount of spring-back occurs when the pressure is released, but this nearly always happens in the first 48 hours and does not affect the overall stability of the bales. Most bales which are produced in this country are of medium density and are not self-sustaining, requiring banding to retain their shape.

Composting

This process is not popular in the UK, but is used extensively in hot climates. The composting process requires a warm climate to be efficient, and results in a material which is used as a soil conditioner rather than a fertilizer. In fact, compost which is made from refuse alone can be deficient in nitrogen, and this imbalance can be re-dressed by the addition of sewage sludge. As a soil conditioner, compost can improve the drainage of heavy soils and improve water retention in light sandy soils.

The composting process is aerobic, i.e. achieved through biological breakdown in the presence of oxygen, and therefore achieves fairly high temperatures (45–65°C) which, if maintained for 24 hours, are sufficient to ensure the elimination of a large number of pathogenic organisms.

Typically the waste (after pulverization in this country) is 'windrowed', i.e. placed in long heaps up to 3 m high, and regularly turned to ensure the introduction of oxygen. After about three months of windrowing the material will have degraded to a stable humus-like substance which can then be screened to remove large inorganic objects and finally put to use.

Various attempts have been made to compost wastes in this country, usually after pulverization in the *Dano* process, but none has proved to be commercially viable in the long term.

Incineration

This is the most effective method of reducing the volume of waste, as only 30–40 per cent of the bulk remains for disposal. It is also the most costly, and its popularity appears to be cyclic. In the 1930s it was a popular method for the treatment of municipal wastes. However, with up to 30 per cent down-time it became an extremely expensive process, and the popularity declined. Now, however, with an increasing shortage of suitable waste disposal sites (particularly in the South East) and the increasing cost of land disposal, it is again becoming a viable method.

Incineration is regarded as a suitable treatment method for household and commercial wastes and for hazardous wastes. Although both types of waste require combustion at high temperatures in the presence of oxygen, the actual parameters are somewhat different. Domestic and commercial wastes require a residence period of 30 secs at a temperature of 800°C.

Hazardous wastes require a residence period of 1 sec at 1000–1100°C or 2 secs at 1200°C in the case of PCBs.

Resource Recovery

The average composition of household waste is given in Table II, and various constituents (including heat) can be recovered from the waste streams. Recently there has been significant pressure to recover and re-use the various constituents of waste. This has arisen (a) from environmental lobbyists who wish to minimize the use of mineral and other finite resources, and (b) from the fact that in certain parts of the country WDAs are exploring any avenue which will prolong the life of traditional disposal routes such as landfill. The WDAs are therefore promoting any exercise which will reduce the volume of waste for disposal.

Once the various constituents of waste are mixed in a dustbin it becomes expensive to separate out the materials which can be recycled. There was, in Doncaster from 1980–86, the most complete plant in the UK for separating out such materials after they have been mixed. This trial plant could separate out tin, aluminium, glass (and separate this into colours) and ceramics. However, it is much easier to recover materials if they

Table II. Composition of average household waste in UK

	% Wet weight
*Paper	31
Vegetable/putrescible	10
*Ferrous metals	7
*Glass	9
*Non-ferrous metals	<1
*Textiles	2
Rubber/leather/wood	4
*Plastics	3
Moisture	25
Fines/dust/ash	8

**These constituents are potentially recoverable*

are kept separate by the consumer/producer, i.e. if glass, tin, etc are never mixed but are disposed of by their own separate routes.

(i) *Glass.* Bottle banks are now a familiar sight in many parts of the country. Sponsored jointly by the British Glass Federation and the WDAs, the bottle banks save the British Glass Federation thousands of tonnes of oil per annum, and also save the WDAs a proportion of their waste disposal costs.

(ii) *Cans.* During the last 10 years there have been several schemes to recycle tin cans. Some have been aimed at recovering tin and steel, some at recovering aluminium; however, not all have been successful. The most notable recent ones have been *Save-a-Can, Cash-a Can* and *Vend-a-Can.* As may be anticipated, the last two operate on a cash-back system.

(iii) *Paper.* Recovery of paper, which is a fluctuating fashion, became popular in the early seventies when environmental consciousness was first being raised. About 17 trees are required to produce 1 tonne of paper and, although this is not a finite resource, people become concerned about the felling of woodlands and forests and the destruction of natural habitats. In the UK WDAs would collect waste paper (i.e. bundles of newspapers etc) separately for recycling. However, in 1974 there was a recession in this field, and some councils were literally left with fields of bundles of waste paper awaiting re-use. For about a decade recycling of used waste paper (as opposed to clean factory waste paper) continued slowly. It is now becoming more popular, and some WDAs are again collecting waste paper and board separately.

(iv) *Plastics.* From the packaging of nearly all consumer products, it would appear that there is a large amount of plastics and polyethylene products which is available for recycling. Unfortunately, these products are made by a variety of processes which are not always compatible with each other. Therefore not only do these materials have to be discarded separately from other waste, they have to be separated out according to the original manufacturing process. This tends to discourage their collection. However, there is a process for collecting and recycling polyethylene tetrafluoride containers, i.e. those containing carbonated drinks, and some supermarkets provide receptacles for such containers.

Refuse-Derived Fuel

The production of refuse-derived fuel (RDF) is more a method of recovering resources from waste than of reducing the volume of waste for disposal. Only about 40 per cent of the waste is suitable for RDF production and there is always a residue which requires disposal. RDF refers to the use of the organic constituents of waste and plastics for the production of energy, by burning. Therefore prior to its use, some sort of separation process, such as shredding or pulverization, must occur.

During the last 10–15 years, research into the use of RDF has progressed significantly and, although there are still many problems associated with its use, its current viability in the UK is uncertain. It is now possible to burn loose pulverized or crushed waste on a coal-type moving grate, with or without coal. Alternatively the waste can be made into briquettes or pellets which can be stored satisfactorily and burned, again with or without coal. Methods are now available for enhancing the calorific value of RDF (untreated at about two-thirds that of coal), one of which is by the addition of an oil—itself derived from waste.

TREATMENT OF LIQUID WASTES

The majority of liquid (or slurry) wastes are hazardous and/or toxic, and generally fall into the 'difficult to handle' category. The treatment of liquid wastes is therefore carried out principally to make them acceptable for conventional disposal. This results in a reduction in their toxic effect and, generally, in their transformation into slurry or cake, which is transferred for instance to a landfill site, and a non-toxic liquid effluent which can be discharged to a sewer.

The treatment of liquid wastes may be carried out by three basic methods: biological, physical and chemical.

Biological Treatment

This process is applicable to large volumes of waste whose toxic content is low, or from which the toxic element can easily be removed. The method relies upon bacteria and other biological organisms to remove organic pollutants from an aqueous waste stream, using their metabolism to change them to innocuous end-products. Treatment processes are similar to those which are employed for sewage treatment: preliminary settlement of insoluble solids, use of a biological media to digest (aerobically or anaerobically) the organic content of the waste, final settlement and reduction of water content of the settled solids.

Depending upon the efficiency of the biological process, the aqueous effluent can be discharged directly to sewer, although it may require 'polishing' by chemical treatment. The sludge can be disposed of by soil injection, landfill, incineration or dumping at sea.

If a waste stream is amenable to biological treatment and there is sufficient capacity at the local sewage-treatment works, the water authority may permit direct discharge of the entire waste stream to sewer. Whether the entire waste stream or the treated effluent is discharged, a licence to discharge will be required. This licence will set limits on volume and the concentrations of certain contaminants.

It should be noted that biological treatment methods are not amenable to shock loadings of contaminants, therefore the waste stream requires careful monitoring to safeguard against such an event.

Physical Treatment

Physical treatment is directed towards the separation of the solid phase from the liquid phase. This includes such processes as screening and filter pressing, and may apply to the raw or treated waste stream. For the purposes of this booklet, solidification or encapsulation is considered with stabilization as chemical treatment.

Chemical Treatment

Although there are many forms of chemical treatment, they are all directed towards reducing the harmful effect of the waste. The principle methods are neutralization, thermal destruction and stabilization.

(i) *Neutralization.* Both acid and alkali waste streams can be neutralized, sometimes by mixing them together. As acidic effluents generally result from metal-plating processes they contain metals in solution. On neutralization, the metal hydroxides are precipitated out of solution and can then be coagulated and separated from the aqueous liquor.

Cyanide wastes are treated by oxidation which transforms the cyanide into an innocuous form. Oxidation is carried out by the addition of an oxidizing agent such as ozone. However, further treatment of cyanide wastes is generally required as they usually contain heavy metals which are often complexed with the cyanide.

(ii) *Thermal Destruction.* Incineration is commonly used for the treatment/destruction of toxic wastes, both at the site of production (i.e. the factory) and commercially on a national scale.

Wastes or effluents having an organic content can readily be incinerated as the organic fraction is easily burnt and may be self-supporting, although some supplementary fuel may be required at start-up and to burn the inert residues. Liquid toxic wastes are more easily incinerated than solid toxic wastes, the latter being mainly drummed resins and contaminated dusts and soils. Most of the toxic or hazardous waste incinerators in the UK are suitable for liquid wastes alone. Four plants are available on a commercial basis: the Rechem Environmental Services Ltd plants at Pontypool and Fawley, the Cleanaway Ltd plant at Ellesmere Port, and the Leigh Environmental Ltd plant at Killamarsh, Derbyshire. There are, however, plans by a number of waste disposal contractors, including Cleanaway and Ocean Environmental, to build new hazardous-waste incinerators in the near future.

The other basic division in type of waste for incineration (other than solid or liquid) is whether it is chlorinated or non-chlorinated. When burnt, chlorinated wastes produce hydrogen chloride gas which attacks linings within the plant and combines with water in the atmosphere to form hydrochloric acid. The requirement to remove this gas, generally by passing through water scrubbers, makes incinerators burning chlorinated wastes more complicated than those taking only non-chlorinated wastes. There is a further complication in that the waste from the water scrubber is then a contaminated effluent, and should be treated as such, and may not always be discharged direct to a sewer.

Polychlorinated biphenyl (PCB) wastes may only be disposed of by incineration in the UK. Partial combustion, i.e. oxidation, of these wastes results in the formation of dioxins and dibenzofurans, both of which are extremely toxic substances. For this reason, PCB wastes may only be burnt in special incinerators where the temperature is maintained at 1200°C with a residence period of 2 secs with 2 per cent excess oxygen. The only plants in the UK which are suitable for burning PCB wastes are the Rechem plant at Pontypool and the Cleanaway plant at Ellesmere Port.

Whilst incineration is the process of thermal destruction in the presence of oxygen, there is another process called 'pyrolysis' which is carried out in the absence of oxygen. Pyrolysis is only suitable for organic wastes, but results in fuel by-products (a gas, an oil and a tarry residue) which can then be used as supplementary fuel in another process. It is not used at present in the UK for toxic waste treatment.

Toxic wastes may be incinerated at sea. As the plant used at sea is less sophisticated than on land, there is some restriction on the type of wastes for which this form of treatment is permitted. PCBs and other chlorinated wastes may not be burnt at sea. However, it is a useful route for a difficult waste known as 'acid tar'.

(iii) *Stabilization.* This classification covers the processes of stabilization, solidification and encapsulation. Although they are different processes, they all result in a slurried cement mix containing toxic waste which is designed to harden and contain the toxic elements of the waste. Strictly, the stabilization process involves the chemical bonding of the toxic elements, such as heavy metals, i.e. they cannot be leached out, for instance, through rainfall passing through the matrix. Encapsulation and solidification involve the physical trapping of the waste which means that, as the hardened solid breaks down, the toxic elements can be released into the leachate.

Two cement-mix processes have been operated in the UK with varying degrees of success. One process, the *Sealosafe* process is marketed in the UK under the trade name *Stablex,* and the other process is known as the *Chemfix* process. Both names imply that treatment is achieved by chemical bonding within the hydrated cement matrix. Whilst this is true for some pollutants, it is not true for all the waste streams for which the processes have been used. The *Stablex* process is operated, under licence, by Cory Waste Management in Essex and by Leigh Interests in the Midlands. The *Chemfix* process was formerly operated under licence by Wimpey Laboratories, and is not used at present in the UK.

The process of hardening or solidification is achieved by various chemical additions to the basic cement-waste mix. The exact proportions and identity of the added chemicals will depend on the waste stream to be treated. The basic *Chemfix* mix consists of cement, waste and sodium silicate. The basic *Sealosafe* mix consists of cement, pulverized fuel ash (PFA) and a pretreated, alkaline waste. Cory uses the process as a partial treatment prior to co-disposal of the slurry in domestic waste. The domestic waste therefore provides an attenuation zone for any substances which are leached out of the mix, and this application has been reasonably successful.

Leigh Interests have used the *Sealosafe* process as a treatment for toxic wastes prior to discharging them down a mineshaft. As with the Cory application, this proved to be acceptable only because any leachates were contained within the mine system.

These processes have occasionally failed in the past and are, as used at present, unsuitable for the treatment of organic wastes, but they have commercial potential. Extensive research is being carried out at Imperial College, London and at Harwell into the mechanisms of encapsulation and stabilization, and into ways in which the processes can be improved to treat a wider range of wastes.

Treatment of Sewage Sludge

The principle methods employed for the treatment of sewage sludge are (i) reduction of water content and (ii) incineration.

(i) *Reduction of Water Content.* Sewage sludge is normally treated to reduce its water content by a combination of 'thickening' and 'pressing', or 'filtering' prior to final disposal. Thickening may be carried out generally by gravity and sometimes following the addition of chemicals. The solids are then separated from the supernatant liquid using a batch process called 'filter pressing', i.e. by feeding the mixture into a plate press under hydraulic pressure; the water being forced through the fabric between the press plates. A cake containing 30–40 per cent solids can be achieved by this method.

An alternative to pressing is by continuous filtering, the most common methods in use being the belt or drum filters. Essentially, thickened sludge is forced under vacuum against a filter, the water passing through, and the solids remaining on the filter. A cake of 18–25 per cent solids can be achieved.

(ii) *Incineration.* Thermal destruction of sewage sludge by incineration is used as a treatment method for about four per cent of the UK arisings. Until recently it has been unpopular because the high costs of fuel have made it unattractive compared to other methods of sewage sludge treatment and disposal. An ash residue remains for disposal—normally to land.

Treatment of Agricultural Wastes

A fairly wide range of treatment is available for agricultural wastes: incineration, drying, mechanical separation and biological treatment.

Incineration is a costly method and, as such, is not very attractive at present. It is only suitable for wastes containing a relatively high concentration of total solids such as poultry and some cattle wastes. The incinerated residue may have some value as a fertilizer.

Mechanical separation refers to the separation of the liquid and solid components of the waste, and is generally carried out to facilitate spreading onto agricultural land. The methods employed are similar to those used for sewage sludge dewatering and include vibrating screens, rotary screen presses and belt presses.

Biological processes for the treatment of agricultural waste encompass both aerobic and anaerobic degradation. Aerobic methods include composting, oxidation ditches, lagoons, biological filters and various methods of aeration. Anaerobic methods include septic tanks, lagoons, filters and digestion.

TREATMENT OF RADIOACTIVE WASTES

The method of treatment which is employed for radioactive wastes is dependent upon the level of radioactivity and the ultimate disposal route. In 1982 the Nuclear Industry Radioactive Waste Executive (NIREX) was set up as a result of a government White Paper (*Command No. 8607).* Its responsibility was to develop comprehensive plans for the safe disposal of low and intermediate-level radioactive wastes, and was funded jointly by the Central Electricity Generating Board, British Nuclear Fuels plc, United Kingdom Atomic Energy Authority and the South of Scotland Electricity Board. In November 1985 NIREX became incorporated as a public company (United Kingdom NIREX Ltd), shares being held by the former funding bodies of NIREX, and with the same responsibilities for the safe disposal of low and intermediate-level radioactive wastes. British Nuclear Fuels also dispose of some low-level wastes independently at their own sites.

The management of high-level radioactive wastes is the responsibility of the United Kingdom Atomic Energy Authority and British Nuclear Fuels Limited.

Low-Level Wastes

Low-level wastes comprise operational wastes such as paper, plastics, metal or glass with a trace contamination of radioactivity arising from nuclear facilities and from industrial and medical uses. Some wastes arising from the decommissioning of nuclear, industrial and medical radioactive plants are also classed as low-level waste because of the low level of radioactivity. Less than 30 per cent of low-level wastes arise from decommissioning activities.

Low-level wastes require no shielding during transportation and disposal, and NIREX have developed a range of containers which is suitable for their handling and transportation: 200-litre drums, a 3-m^3 box and 12-m^3 box, the last of these for large items of decommissioning wastes. For transport the 200-litre drums (the 'normal' or most commonly-used containers) will be loaded into an International Standards Organization (ISO) type freight container. Prior to disposal the drums will be compacted, loaded into a standard box and the voids between them filled with concrete.

Intermediate-Level Wastes

Intermediate-level wastes comprise similar types of waste to those described for low-level wastes, but with higher radioactive contamination. They arise principally at nuclear sites. Although some heat may be generated, it is not to such an extent that it has to be taken into account in either its treatment or disposal.

Intermediate-level wastes require shielding, and NIREX have developed a range of containers which is suitable for their handling and transportation. The normal containers are 500-l unshielded drums which, however, are placed in a shielded container for transport and other handling. The 3-m^3 capacity unshielded boxes are also placed in a further shielded container for handling. A 12-m^3 capacity self-shielded box is available for large items of decommissioning waste.

High-Level Wastes

These long-lived heat-generating wastes are not suitable for disposal. The current strategy is that they should be converted into glass blocks and stored for at least 50 years.

TRANSPORTATION OF WASTES

Three main options are available in the UK for the transportation of wastes: road, rail and barge, and the relative uses of these methods are given in Table III.

Table III. Methods of transportation for municipal waste

	Per cent
Road	94
Rail	4
Barge	2

Road Transport

The majority of wastes are transported by road, about 70 per cent being transported directly to the waste disposal site in the initial collection vehicle. This may be in municipal collection vehicles (containing mainly household waste), tipper lorries or skips (containing commercial and dry industrial wastes), or tankers (containing liquid wastes).

Of the wastes which pass through a transfer or treatment station the majority is again transported by road. However, in this case the waste will generally have been packed into larger containers, where possible increasing the bulk density, to economize on transportation costs. Special vehicles are available for the bulk transportation of large volumes of waste.

Road transport is the most commonly-used method because waste tends to arise from discrete communities and may then be sent to disparate treatment or disposal sites, depending on its character and the capacity of the facilities which are available. It is common for a waste disposal authority to transport municipal waste from different parts of the town to different landfill sites, and for industrial and commercial wastes to be taken to different sites. Alternatively, municipal or industrial wastes may undergo a treatment process prior to final disposal.

Rail Transport

Rail transport may be used when (a) there is a good rail link between the waste source and the disposal or treatment site, and (b) large quantities of waste are generated at one place and the treatment or disposal site has capacity for this volume. It is not generally economic to transport less than 600 tonnes per train load. At an average capacity of 20 tonnes per ISO container this amounts to about 30 containers at one time. Municipal waste in the UK is generated at an average rate of a third of a tonne per person per year. Therefore to fill a train load every week a minimum population of 100 000 is required as a waste source.

Several towns and cities now use rail as a transportation method for their municipal waste, the most notable being Manchester and London. Generally, a loading terminal is built at the transfer station to which, if necessary, waste will be transported by road from other transfer stations to make up the load. An unloading terminal must also be constructed at or near to the disposal site. As this is costly, it is generally only carried out either at large capacity landfill sites or where potential disposal sites are close together, making the final road transfer feasible and economical. A former minerals loading rail facility may be in a suitable location to be refurbished for this use.

Barge Transport

Barge transport of waste has similar requirements to rail transport to make it economical: a large volume at one source, a large capacity site, and a water link. Because of this last requirement it is of far more limited suitability than rail transport, and has only been used successfully to transport London's waste to the Essex marshes disposal sites. As for rail transfer, the waste is loaded into ISO containers. It is then lifted onto and off the barges by special crane facilities constructed on the docks. At the landfill site the containers have to be transported by special lorries to the disposal point. The practice of loading the wastes directly onto open barges for subsequent unloading by grab crane at the disposal site has virtually ceased. It was felt to be most unsanitary, the waste being available to vermin and also open to rainfall, making it more difficult to handle on unloading.

4. DISPOSAL

INTRODUCTION

In this booklet the term 'disposal' is taken to mean the final resting place, and as such there are only two options, i.e. to land or to sea. The majority (as much as 97 per cent) of all wastes which are generated in the UK are disposed of ultimately to landfill, the remainder being dumped at sea.

The storage of wastes underground, i.e. in mineshafts and caverns, is considered separately from landfill (although in this section) as it is not generally accepted as a final disposal method. Incineration of wastes does not usually result in total destruction, the residue requiring disposal in a landfill, and is therefore considered in Section 3.

LANDFILL

The landfill disposal of wastes has become a science, growing out of the American practice of sanitary landfill which replaced open tipping in the UK from about the 1950s. There are still many badly-operated sites, and many which have not been properly engineered; however, standards are generally improving, and waste disposal is becoming much safer and more acceptable.

General guidelines for the safe disposal of wastes in a landfill site are given in the DoE's *Waste Management Papers 4* (as republished 1988) and *26* (published 1987). These can be summarized as follows:

(i) The site should be geographically or geologically located such that polluted water from the area cannot contaminate potentially usable aquifers, or otherwise an engineered liner should be provided;

(ii) Wastes should be placed in layers generally 1–2 m deep and compacted;

(iii) Inert cover should be provided to the wastes on a daily basis;

(iv) Litter and pests should be controlled;

(v) The site should be covered and restored to a use which is compatible with its nominated land use and its surroundings; and

(vi) Provision should be made for the monitoring and, if necessary, control of landfill gas and for the collection and treatment of leachate prior to discharge to a sewer or natural watercourse.

Types of Sites

Landfill disposal of wastes has been used, as the name suggests, to fill natural or man-made depressions in the landscape and to reclaim marginal land. In Europe and in some rare instances in the UK a different kind of disposal to land occurs, i.e. 'landform', where waste is placed on top of a raised landscape to raise it further.

Typical sites for landfill are quarries resulting from mineral extraction, valleys and depressions and disused railway cuttings. Marginal marshland can be raised and reclaimed using waste as fill material. Land can even be reclaimed from the sea, although this is not common practice in the UK.

Two methods of 'best landfill practice' have previously been acceptable in the UK: one favoured the 'dilute and disperse' philosophy, the other 'concentrate and contain'. The terms related primarily to the attitude towards leachate—the polluted water coming out of solid waste. The 'dilute and disperse' method applies only to sites in permeable strata. It is assumed that there is a sufficient depth of rock (known as the attenuation zone) between the base of the landfill and any potentially-usable aquifer which is capable of absorbing the polluting nature of the leachate. This basic philosophy for landfill site design is no longer favoured in the UK, and can only be adopted in special cases where the geology is well documented and there is little likelihood of a pathway developing to a potentially-usable aquifer, the site itself not lying over such an aquifer. A further area of concern arising out of the use of these sites is that the geology is also permeable to gas migration, the hazards of which are outlined later.

The 'concentrate and contain' philosophy assumes that no water within the site will be allowed to leave the area through the ground either below or surrounding it. This gives maximum protection to all groundwater in the area, and is the favoured practice in the UK. Leachate is collected within the site and is pumped out for treatment and discharge. Gas may be effectively dealt with in a similar manner. If the geology is not such as to provide a sealed site, this must be provided by the preparatory engineering works.

Planning

Planning of a landfill site begins long before a planning application is lodged with the appropriate planning authority, i.e. county in England, district in Scotland and Wales (see section 2).

Whether it is a waste disposal authority or a private contractor who is seeking a disposal site, the planning and feasibility stages will be similar. The first task is to identify a suitable site to which waste can reasonably be transported by road, rail or barge. In parallel with this, the potential source(s) of waste must be identified in terms of type, volume and rate of waste generated, methods of transport available and pretreatment, if any is carried out.

It will then be necessary to examine the site to assess its suitability in terms of geology, volume, proximity to a suitable watercourse for discharge of treated leachate, and proximity to a suitable transport terminal. The preliminary financial viability of the operation will also be assessed. Discussion with the water authority (or national river authority) usually takes place at this stage to ascertain whether or not a particularly sensitive aquifer lies in the proximity of the potential site.

If this preliminary screening process shows that a particular site may be appropriate for the disposal of an identified waste stream, the planning authority may be approached, on an informal basis, for their first views on the proposal. This allows the planning authority an opportunity to object to a proposal before expensive site investigations and preliminary engineering design are carried out.

Guidelines for the site investigation are given in detail in *Waste Management Papers 26 and 27. Waste Management Paper 26* gives guidelines for a 'detailed site investigation' in section 3.109. In summary, this includes geological and hydrogeological investigations and a topographical survey of the site. The site investigation report should identify features and give sufficient information for the following to be evaluated: site preparation work, water pollution prevention measures, pollution monitoring, gas venting and control requirements, intermediate cover requirements, quantities of top soil, sub-soil and soil-making material available on site, estimated quantities of soil needed (to be imported for restoration) and vehicle and machinery requirements.

Waste Management Paper 27 includes a survey of all man-made services and natural (i.e. geological) discontinuities around the site for a distance of 250 metres. This survey is intended to locate all (or as many as possible) of the potential gas migration pathways around the site. The remedial measures and preparatory works which are identified as necessary from this site investigation allow a more realistic assessment to be made of the costs of the proposed landfill operation.

Planning applications are supported by the results of the site investigations, together with an assessment of the possible impacts on the environment, amenity and, where relevant, agriculture. An accompanying statement and drawings should also demonstrate preparatory engineering works and the method by which the disposal operation will be carried out, together with details of the final restoration of the site. Neighbouring landowners must be notified of the application by notices posted on the site and in local newspapers, with the full application being available for view in a public place (school, library) local to the site.

Engineering Design

Engineering design of a landfill site encompasses all the preparatory works which are required to make the site suitable and safe for the disposal of wastes, the necessary accompanying works, the detailed considerations of the phasing of the waste disposal operation, and the final restoration.

Preparation works will include the design of such items as site liners, a leachate-collection system, cell construction, haul-road construction, provision and suitable location for storage of daily cover material. Site liners may be formed out of natural materials such as clay which is engineered to a specified permeability limit, or impermeable or low-permeability synthetic materials. *Waste Management Paper 26* gives ample guidance on the variety of liner methods which is available.

Accompanying works include the construction of a (generally concrete) site access road, site fencing and notice boards, provision of a wheel washer, a weighbridge (if required) and site offices and laboratory (which is required where hazardous, toxic or special wastes will be coming onto the site).

Phasing of the waste disposal operation will have to take account of the rate of waste input to the site, the overall water balance and the practical size for cell construction. The ideal situation is generally to minimize water ingress to the waste (and thereby leachate production) by keeping to a minimum the surface of waste which is exposed to rainfall.

Engineering design of the restoration phase includes consideration of the final contours to encourage surface-water run-off, provision of a surface run-off collection system, inclusion of adequate capping material (such as clay), subsoil and topsoil for the proposed end use, and a planting scheme for the final restored surface. Filling to achieve the proposed contours must take into account settlement of the waste due to compaction and decomposition, and the differential effect of waste settling against the boundary of the site and at the cell walls.

Sundry other works include the provision of sight and noise barriers to protect local residents from disturbance during the operation, and the provision of a landfill gas monitoring and collection system. Gas monitoring (see *Waste Management Paper 27)* is generally carried out from boreholes. This may require the construction of an inert layer around the waste, within the site, if access cannot be gained to land outside the site to place boreholes. The permission of the relevant landowner is required for the drilling of boreholes.

Leachate

Leachate is water which has come into contact with refuse and contains contaminants from that refuse in suspension or solution. Sources of this water are rainfall, groundwater infiltration, water contained within the refuse itself (typically 20–25 per cent in domestic refuse), water from liquid waste disposal and water resulting from the biological decomposition of organic wastes (Fig. 4).

Contaminants which are present in leachate are typically metals, organic acids and esters, ammonia and other nitrogenous compounds. Pesticides and PCBs may also be present. Leachate generally has a high chemical oxygen demand and low oxygen content although, as with all the parameters mentioned here, these will vary widely from site to site and with the age of waste within the site.

On a 'dilute and disperse' site the leachate is allowed to percolate out of the site into the rock below, relying on the physical entrapment of the solids and the chemical alteration of contaminants in solution to prevent serious groundwater pollution.

For a 'concentrate and contain' site it is necessary to collect the leachate for treatment prior to discharge away from the site to a suitable watercourse or sewer.

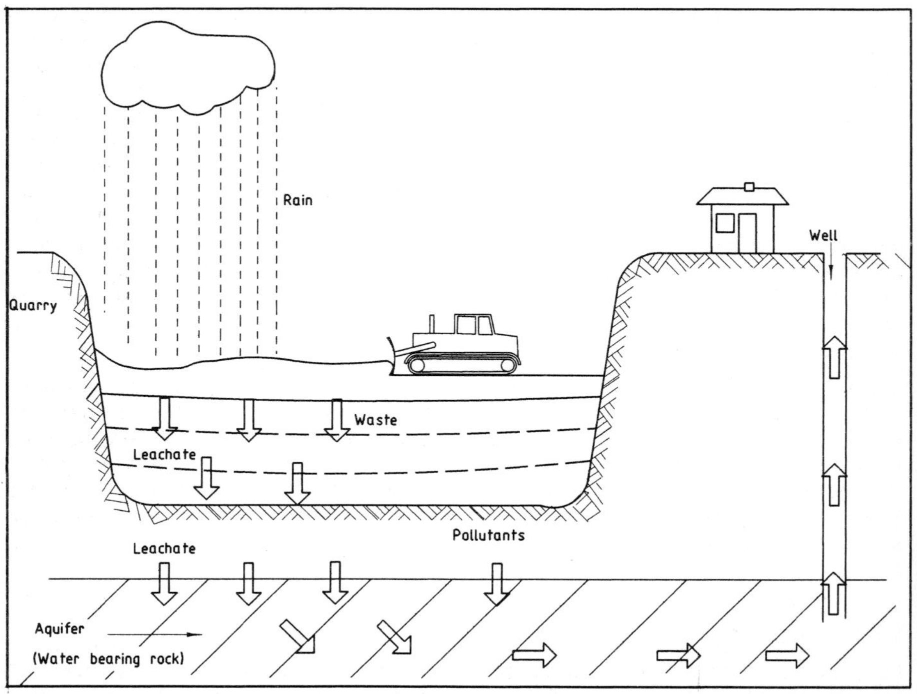

Fig. 4 Environmental impact: groundwater pollution

Collection of the leachate can be undertaken from the base of the landfill site with a herringbone-type structure of collection pipes; alternatively a collection system can be installed around the site a few metres below the lowest point on the rim.

Discharge of leachate to a watercourse, whether it be a river/stream or a sewer, requires a discharge consent under the Control of Pollution Act 1974. If granted, a consent will set upper limits for certain contaminants anticipated to be present in the leachate, which may prove harmful to the watercourse. If a consent is granted for discharge to a sewer, it will be treated as an industrial effluent. In this case the water authority will make a charge for the discharge to cover the cost of treatment at its own works; the charge will be based on a formula which considers such aspects as volume, chemical oxygen demand, suspended solids and may also include multiples of other parameters.

Whether the discharge is to a river/stream or to a sewer it is generally necessary to treat the leachate prior to discharge; firstly to achieve the consent limit, and secondly to minimize the charge which is payable. As ammonia is the major contaminant in leachate and the potentially most harmful substance to a river/stream (kills fish) or a sewer (harmful to biological treatment processes at sewage-treatment works), most leachate treatment is aimed at reducing ammonia and chemical oxygen demand. The most frequently-used treatment method is that of oxidation. This can be achieved by the addition of chemicals, such as hydrogen peroxide, or by aeration. Combinations of the two methods have been used with varied success, as have biological methods such as activated sludge or biological filters/contactors. With biological methods it is generally necessary to add nutrients such as phosphorus or phosphoric acid, to enable the biomass to thrive. Recirculation of leachate through the waste as a biological reactor is thought to reduce the strength of some leachates, and can be achieved by spraying over the surface or by direct injection into the waste. Direct dilution with clean water is another possible method.

As treatment of any form is an additional cost in the operation of the site, good practice now centres on reducing the volume of leachate (and the volume to be treated) as far as possible. This requires operation of the site in such a manner that all clean (i.e. not been in contact with waste) rainfall run-off and groundwater seepages are collected separately from the leachate, and that the surface of waste exposed to (and able to absorb) rainfall is kept to a minimum. Clean waters can be discharged with little (possibly settlement alone) or no treatment to a stream/river or allowed to soak away into the ground.

Landfill Gas

Landfill gas comprises methane (60 per cent) and carbon dioxide (40 per cent), and is produced as a result of the decomposition of the organic matter which is contained in (domestic) waste. Methane is flammable in air at a range of 5–15 per cent concentration by volume, provided that an ignition source and sufficient oxygen are present. Carbon dioxide has asphyxiating properties and is normally present in air at 0.03 per cent by volume. Guidelines in *Waste Management Paper 27* give safe levels within a dwelling: less than 1 per cent for methane, and less than 0.5 per cent for carbon dioxide. Above these levels action, such as active venting and/or evacuation, should be taken.

During recent years, landfill gas has become a matter of public concern because of a number of 'incidents' (explosions), the most serious of which occurred in Loscoe, Derbyshire in 1986. No-one has yet been killed as a result of one of these explosions, but there have been injuries. This has prompted the DoE to publish the guidelines in *Waste Management Paper 27* and to sponsor research to provide further knowledge of the gas.

The problems relating to landfill gas have arisen from two main causes: the first is the way in which waste is landfilled, and the second is the increasing re-use of derelict land for redevelopment.

Placing biodegradable waste well mixed, in a sealed landfill site, adding an optimum

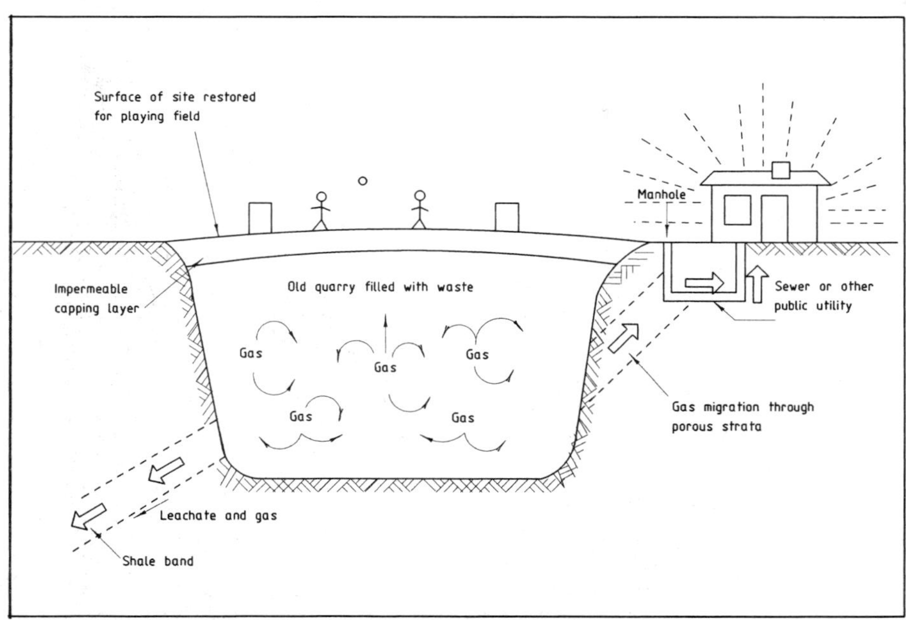

Fig. 5 Environmental impact: gas

Plate 1. Rear loading mobile compaction vehicle

Plate 2. Waste collection unit for recycling

Plate 3. Barge transfer of waste

Plate 4. Landfill site near Poole, Dorset

Plate 5. **Typical landfill compactor**

Plate 6. Leachate treatment plant at Compton Bassett, Wiltshire

Plate 7.   Transfer of liquid hazardous waste

Plate 8. Hazardous waste incineration

amount of water, sealing daily and finally placing an impermeable cover creates ideal conditions for the production of landfill gas (Fig. 5). If no measures are taken to remove this gas by venting or flaring, pressure can build up to a level which encourages the gas to migrate along easy pathways such as natural rock fissures or man-made services. The gas can thus make its way off site (migration distances of up to 400 m have been recorded) and into buildings where electric sparks and cigarettes can provide the ignition source required to cause an explosion.

The re-use of derelict land for redevelopment has led to considerable development on or near to previously-closed landfill sites. These sites are unlikely to have had gas-control measures installed, and development adjacent to such a site should be treated with caution. The Interdepartmental Committee for the Redevelopment of Contaminated land (within the DoE) issue guidelines for the type of site investigation which should be carried out prior to the development of any derelict land site. This includes a survey to establish whether methane gas is present on the site — whether residual, being generated, or migrating on to it.

Most planning authorities now require consideration to be given to the monitoring and control of landfill gas, at the planning application stage. Although it is not strictly a planning or land-use issue, it does affect the potential use of that land or of adjacent land. There will almost certainly be site licence conditions imposed requiring the gas to be monitored, and controlled if necessary.

Guidelines are given in *Waste Management Paper 27* for both the monitoring and control of landfill gas. Its commercial use is considered in *Waste Management Paper 26.*

In summary, monitoring should take place at the edge of the landfill site, at a point approximately half way between the edge of the fill material and the nearest property at risk, and adjacent to the property. If no such property exists within 500 m of the site, then monitoring should be carried out at 250 m and 500 m from the fill materials. Although not precisely stipulated, it is generally accepted that monitoring points should be boreholes sunk to the approximate depth of the landfill.

Control measures will depend upon the particular geology surrounding the site and the proximity of local development (present and proposed). An impermeable barrier may be required to prevent gas migration off site. Adjacent to this should be placed a venting ditch, for example filled with stone to give a large void ratio. It may be sufficient to allow this ditch to vent direct to atmosphere, in which case measures must be taken to prevent soil and plants sealing the surface. If, however, this is insufficient, a gas-collection system may be installed. This may be placed as part of the venting ditch, or as a herring-bone lateral and chimney system, or a combination of both. Design of these systems should take into consideration the final restoration and proposed end-use of the site; for example, ploughing activities should not be capable of ripping up the gas-collection system.

Landfill gas may be used as a source of fuel. The hindrance to its use is generally the difficulty of finding a potential user within economic/feasible distance of the site. However, there are now many instances in the UK where landfill gas is used commercially, and several where it has been converted to electricity. The Government has announced proposals for 1992 which should encourage the further use of landfill gas. The Energy Technology Support Unit (ETSU) at Harwell can give guidance on the commercial use of landfill gas and on the grants which are available to assist such schemes.

Reinstatement

Landfill sites can be restored to suit three types of end-use: (a) recreation, (b) agriculture, and (c) 'hard' development (building), and this should be taken into consideration at the planning stage. Generally in the UK, hard development is not encouraged as a proposed end-use for a potential landfill site, although it is being increasingly considered for closed landfill sites (see previous section).

Recreational use is a favourable after-use in an urban or semi-urban area. This use requires the site to be filled so that the final settlement is fairly even to prevent, for example, ponding occurring on a sports field. However, it does not require a great depth of sub and top-soil over the capping layer, as the principal vegetation is grass.

Agricultural after-use is preferred in rural areas. Landfilling frequently occurs through the infill of valleys leased from a farmer who requires them to be returned for his use on completion. In this case a small amount of differential settlement will not be quite as important as for a sports field, provided that there is adequate drainage. However, a good depth of sub and top-soil is required to prevent the capping layer from being ploughed up and damaged. Also, an increased depth will be required where trees or hedges are to be planted.

Waste Management Paper 26 gives guidelines for the depth of capping layer, and sub and top-soils required. These should be treated as minimum depths, and discussions should be held with the planning authority as to their general requirements.

General practice is to place an impermeable layer, such as clay, over the landfill site prior to the placing of sub or top-soil. A suitable thickness is considered to be 0.75–1.00 m, to prevent ingress of rain-water to the site. However, clay will crack if it is allowed to dry out, permitting both water ingress and gas egress. Evidence of the cracks is usually given by vegetation die-back where the accumulation of gas has starved roots of oxygen, killing the plants.

Various alternatives to this practice have been tried, and research is currently being carried out to monitor the efficacy of some of these alternatives. Some possibilities include (a) the use of a permeable layer (to collect the gas) below the clay layer, (b) a combination of natural materials to provide a flexible impermeable layer which will not crack, or (c) the use of a combination of synthetic and natural materials (such as an impermeable synthetic layer underlying a drainage layer).

Obviously the design of the final capping layer and restoration surface will have to take account of such items as the gas control system, and the need to keep a venting ditch open to the atmosphere. If leachate recirculation by injection is being considered as a method of treatment, provision will have to be made for this below the capping layer.

Because the capping layer is generally impermeable or of low permeability, the restoration contours have to be designed to encourage run-off. To prevent this run-off entering the waste at the edge of the site, it is usual to install some sort of collection system at the site perimeter. This should be kept totally separate from the leachate system, and may be discharged to a watercourse or into a soakaway system remote from the site.

Types of Wastes

The types of waste which may be placed in a landfill site are limited by the conditions of the site licence. Most waste disposal authorities have a range of categories of waste, and will normally limit a site licence to certain of these categories, with certain specified exclusions. A limit may also be placed on tonnages of any specified wastes over certain time periods.

Typical categories of waste are:

(i) 'Inert' materials, generally including building materials but not asbestos;

(ii) Materials of 'low biodegradability' such as non-toxic industrial wastes containing some organic matter, for example paper and plastic wastes, road gully washings;

(iii) 'Biodegradable' wastes including household waste;

(iv) 'Difficult' wastes, which are those requiring special handling but which are not necessarily toxic or hazardous; and

(v) 'Hazardous' wastes which include toxic, hazardous and special wastes.

The type of waste which is allowed in a certain site will be dependent on the type of site, i.e. containment or dispersal, and on the sensitivity of the site in terms of local opposition or sites of special interest/For instance, unlined sites in chalk, overlying currently or potentially-usable aquifers may only be permitted to take the inert category of waste. Alternatively a site in low permeability clay remote from any aquifer may be permitted to take difficult and hazardous wastes. The treatment of such wastes (see later) may make them more acceptable for disposal at a greater number of sites.

Most wastes which are disposed of in a landfill site are solid in nature. However, with careful attention to the overall water balance, sludges, silts and slurries may also be disposed of in this way. Liquid hazardous wastes may be placed in a landfill site by a method known as co-disposal.

Co-disposal

Co-disposal is a term which is used when two types of waste, generally a solid and a liquid, are disposed of together in one site.

Two techniques are employed for the actual placing of the waste. The first involves placing the two types of waste in alternate layers with daily cover on top of both. This method may be appropriate for silts and sludges, provided that it does not present handling difficulties.

The second method requires the prior disposal of solid, absorbent waste by the normal landfilling method—in layers with inert daily cover. Trenches are then dug into this material and the liquid, slurry or sludge is poured into this hole. The waste, which is already in place, absorbs the liquid and acts as an attenuation zone for gross contaminants if the liquid is hazardous or toxic.

Considerable research has been carried out, mainly by the Water Research Centre, into the co-disposal of various combinations of waste.

Treatment Prior to Disposal

Section 3 considers all kinds of waste treatment from simple compaction in a transfer station, through the various requirements to facilitate resource recovery, to the treatment of hazardous wastes. Incineration is also discussed.

Treatment of bulk (non-hazardous) wastes prior to landfill reduces the volume requiring disposal, and simple compaction at a transfer station can assist the achievement of good compaction ratios on the landfill site. Pulverization/shredding increases bulk density considerably and further aids final compaction; it also encourages rapid biodegradation with fairly even settlement. Since pulverization and shredding may be used as a first step in the process of resource recovery, the removal of such materials as can be re-used will reduce the volume for both transportation and disposal. This should be taken into account when considering the relative costs of the treatment process compared to the costs for transportation and disposal. However, consideration should also be given to the fact that, with materials such as metals and glass recovered, the remaining wastes have a higher organic content than before. This may lead to the production of a higher strength leachate.

Baling as a treatment process results in wastes which are easy to handle in transportation and on site. However, the compaction ratio which is achieved on site is not significantly different to that which is achieved through prior pulverization or by using good compaction equipment. Water is not easily able to percolate through the waste and biodegradation may, as a result, be inhibited. As baling is still a fairly recent treatment method, little is known of the long-term effects on leachate and gas production.

Treatment by incineration should leave a biologically-inert residue (30–40% by volume of the feed material) which is generally disposed of to landfill. This residue may be high in heavy metal content but, provided that this is not a problem, it may be used in place of or to supplement the inert daily cover material.

DISPOSAL AT SEA

Disposal at sea is reserved for wastes for which a suitable route does not exist on land. Therefore inert materials, demolition materials and domestic waste tend not to go by this route; whereas difficult, toxic and hazardous wastes, sewage sludge and, until recently, radioactive wastes were all disposed of at sea. This method of disposal is considered in the UK to be the best practicable environmental option for certain wastes, particularly in view of our small land area and easy access to the sea. However, the UK has been under pressure from the other EC Member States and from the other members of the North Sea Commission to reduce this practice and consider ceasing it altogether. An agreement was signed in 1987 between the members of the North Sea conference to restrict dumping at sea by 1992 and to attempt to further reduce the quantities after this date.

The disposal of waste at sea comes under the jurisdiction of MAFF, who have an obligation to see that all land-disposal options have been considered and found not to be feasible, before they will issue a licence for sea disposal.

Hazardous Wastes

Hazardous wastes are transported for both incineration at sea and deep sea dumping, and about 7 per cent of hazardous wastes are disposed of by this route.

(i) *Incineration at Sea.* Incineration at sea is subject to certain controls, specifically regulating the temperature of incineration and the degree of combustion of the emission by measuring the percentage of carbon dioxide and oxygen in the exhaust gases.

Marine incineration has been an increasing practice for the UK since 1981 (then 850 tonnes; to 3754 tonnes in 1986) and has been specifically reported on by the Hazardous Waste Inspectorate in all three of their reports (1985, 1986, 1988). The *EEC Community Directive No. 8805/85* calls for a ban on marine incineration; Rule 2(3) of the Protocol amending the *Convention for the Prevention of Marine Pollution by Dumping from Ships and Aircraft* requires a date to be set to terminate the incineration of waste at sea.

For the hazardous waste producer, marine incineration is an attractive option in economic terms, the cost being about one twentieth of that of land incineration (the main alternative).

The main streams of waste which are transported for marine incineration have a high organic content and are generally solvents, acid tars and other contaminated petroleum-based products.

Marine incineration of UK wastes is currently undertaken on board the *Vulcanus II,* a Dutch ship. This ship collects waste for incineration from the various land-based terminals and does not commence incineration until it is out at sea.

(ii) *Dumping at Sea.* Very few wastes are now dumped at sea. The practice, with regard to hazardous wastes, consists of placing the waste in containers sometimes with prior treatment, transporting them to a specified disposal site and then releasing them into the sea at this disposal site. Modified barges are normally used for this exercise, sometimes with a well formed out of the centre so that the drums can be dropped into still water.

MAFF have a certain number of specified marine disposal sites around the UK on the Continental Shelf. Some, but very few, wastes are disposed of off the Continental Shelf in the deeper waters of the North Atlantic.

(iii) *Treatment Prior to Marine Disposal.* Marine-disposal operations are linked to land-based terminals, of which there are twelve in the UK. These terminals are used to store wastes until there is sufficient quantity to make the disposal economic. During this storage period there is opportunity for settlement, e.g. of the solid fraction of acid tars, recovery of solvents etc.

Radioactive Wastes

Marine disposal used to be considered the best environmental option for low and intermediate-level radioactive wastes. A number of sites have been used by the UK, commencing in 1951. Most are to the west of Britain and the EC Member States, although the Hurd Deep in the England Channel was used during the 1950s and early 1960s, and a site north-west of Ireland was used occasionally in the early 1950s. The Holliday report *Report of the Independent Review of Disposal of Radioactive Waste in the North-East Atlantic* (1984) fully describes the UK practice of annually dumping radioactive wastes at sea.

The dumping of solid radioactive waste in the North-East Atlantic was suspended in 1983, and is no longer considered to be an acceptable practice. Such wastes are stored at specified locations in the UK (such as Harwell, Sellafield and Dounreay) until a suitable land-disposal route can be found for them. However, the practice of discharging liquid effluents contaminated with radioactivity is still carried out. The discharge outfall from the nuclear plant at Sellafield has been the subject of frequent attention from Greenpeace and the media.

Sewage Sludge

The disposal of sewage sludge to sea is practiced widely in the UK, over 30 per cent going by this route. Many of the water authorities and regional councils currently dispose of some sludge to sea, and it is probable that the water plcs will continue the practice with little change.

There are eleven locations around the UK coast which have been licensed and used for the disposal of sewage sludge (Table IV). All but the site on the Clyde are rapid dispersion sites, swept by tidal currents.

Table IV. Sewage sludge dumping sites in UK

Firth of Forth (2 sites)	Plymouth
Tyne/Tees	Falmouth (now disused)
Humber	British Channel
Thames (3 sites)	Liverpool Bay
Isle of Wight	Firth of Clyde
Lyme Bay	

Sewage sludge is principally taken by barge to disposal sites licenced by MAFF, little being discharged by pipeline. It is not containerized, and is disposed of by discharging from the bottom of the barge. The sludge may or may not be treated (to reduce the moisture content and hence increase the volume transported at one time) prior to filling the barge.

It seems unlikely that this practice will be reduced significantly in the next few years, although various undertakings have been instructed not to increase the quantities beyond the 1987 level.

DISPOSAL UNDERGROUND

This disposal route is separate from the use of waste to infill quarries and other holes in the ground, and is applied to the deep disposal of wastes in mineshafts, worked-out mines or in specially-engineered deep geological repositories. It is generally used as an intermediate disposal or storage option rather than ultimate disposal.

Deep underground disposal is normally reserved for the disposal or storage of toxic, hazardous or radioactive wastes. There are certain geological formations in the UK which are considered to be fairly secure in that they either have extremely low permeability or are known to be sealed from contact with any groundwater aquifers (for instance, the deep mine shafts at Brownhills in the Midlands, used by Leigh Interests). The intention is that highly dangerous wastes should be securely contained where they can easily be monitored and retrieved if necessary, until by-products can be recovered or a safe disposal route can be found. This is a particularly useful method for wastes which decay and generate heat, such as radioactive wastes.

Radioactive Wastes Underground

NIREX is responsible for managing the safe disposal of low and intermediate-level radioactive wastes. They are currently looking at shallow burial grounds on British Nuclear Fuels Ltd sites for the safe disposal of low-level wastes.

A programme to investigate and provide further near-surface repositories began in 1985–86. The findings of this programme have led NIREX to conclude that 'The economic advantages of near-surface low-level waste disposal were not as great as had been considered earlier. Although near-surface disposal of low-level waste would be safe, the development of a new near-surface repository for low-level waste showed no significant cost advantage when compared with the marginal cost of disposing of low-level waste along with intermediate-level waste in a deep repository, even allowing for the cost of any storage of low-level waste needed before disposal' (NIREX Annual Report 1987–88).

NIREX has also instigated a programme to find deep repositories for low-level and intermediate-level radioactive wastes. Preferred hydrogeological environments have been identified, but specific sites have not been named. Parallel investigations are being carried out at BNFL Sellafield to assess the deep geology below this site and its suitability as a radioactive waste repository.

High-level radioactive wastes, when generated, will be encased in glass (vitrified) and stored in deep repositories where they can be remotely monitored and treated to reduce the heat which is a decay by-product.

MISCELLANEOUS WASTES/ROUTES

Agricultural Wastes

The principal route for the disposal of agricultural wastes is land disposal. The waste may be a liquid, slurry or wet solid and hence may be sprayed, injected or spread and ploughed in as appropriate. A small percentage (e.g. of poultry waste) is recycled through use in the production of feedstock.

Sewage Sludge

Whilst about 22 per cent of sewage sludge is disposed of to landfill, about 45 per cent is disposed of to farmland. It is normal to thicken or dewater sewage sludge which is destined for disposal at a landfill site but, even so, it is difficult to handle in this environment. Disposal to agricultural land does not require prior treatment of the sludge, but is subject to strict controls regarding gross loading of pollutants such as heavy metals. An EC Directive was enforced in June 1989 which limits the application of sewage sludge to agricultural land by stipulating either the gross loading in any one application or the period of time over which such loading may take place.

In the past, methods of application were restricted to spraying, which was cheap and easy. However, the objectionable odours which were released by this method have led to the development of more aesthetically-acceptable techniques such as sub-soil injection of the sludge. This may be carried out by dragging a spray nozzle in the ground on a tine suspended behind a tractor or by driving a mole through the ground whilst injecting.

5. ECONOMICS

INTRODUCTION

The cost of waste disposal includes all activities from collection at the source through transfer, treatment and transportation to ultimate disposal, environmental monitoring and rectification. It therefore includes such items as vehicle and plant costs, maintenance, standing time, labour, fuel, landscaping, planting, monitoring, and discharge costs (i.e. for leachates and contaminated waters to sewer). It also includes some payback for preparatory costs of design and the work required to obtain either planning permission or a licence.

Table V shows typical costs and charges for different parts of this process, calculated for household (municipal) waste. Indicative prices only are given for hazardous or difficult wastes.

OVERALL PROCESS AND DISPOSAL COSTS

The figures given in Table V are taken from the *Waste Disposal Statistics 1988–89 Estimates* published by the CIPFA Statistical Information Service. These figures, both estimates and actuals, are published annually and include waste arisings and costs, including income and expenditure of WDAs. The statistics are compiled from information obtained from the WDAs (83 included out of a total of 106) in England and Wales. Although they are the only waste disposal statistics which are available, they are generally acknowledged within the industry to be deficient in many respects, and should be treated with caution.

It is interesting to note that the estimated total expenditure on waste processing and disposal in England and Wales for 1988–89 was over £216 million, amounting to an overall unit cost of £5.19 per head of population. This unit cost includes employees, running expenses such as administration support, transport, treatment/disposal by other bodies and other sundry operating expenses, capital financial charges and any income derived from the waste, for instance by charging for disposal by others at WDA sites.

Table V. Operating costs (£ per tonne)

	Average	*Range*
Primary treatment		
Transfer—compacting/shredding/baling	4.3	2.2–18.8
Transfer—civic amenity sites	4.5	0.0–19.5
Incineration	11.7	11.6–25.5
Reclamation/recycling	6.3	−6.1–11.8
Final disposal		
Landfill by WDAs	2.8	0.0–15.6
Disposal by contractor's agents and other WDAs	3.5	0.0–24.8

The wide range in the costs given in Table V generally reflects the difference between scattered and concentrated communities and the effect of being unable to place, for example, a landfill site within an urban area. In Wales, rural areas have higher costs of disposal than urban areas; in England the reverse is true. This apparent paradox may be accounted for by the relatively high costs of small-scale operations in Wales and by the need for expensive treatment methods at the large urban conglomerations in England.

Costs given in Table VI are derived from figures given in *Wastes Management Paper No. 26.* The figures are quoted for a range of sizes of sites at mid-1984 prices, and unit costs vary from £2.43/tonne for a large site (250,000 tonnes/annum input, overall volume of 5 Mm^3) to £6.09/tonne for a small site (25,000 tonnes/annum input, overall volume of 0.5 Mm^3). Table VII lists items which should be considered in arriving at these costs.

Table VI. Landfill unit costs for average landfill as given in Table V

	(£)
Site assessment	0.02
Site development	0.23
Operational costs	2.37
Restoration	0.06
After-care	0.09
Unit cost/tonne	£2.77

Table VII. Activities considered in arriving at landfill unit costs

(a) *Site Assessment*
Reconnaissance
Market survey
Preliminary ground investigation
Full geological and hydrogeological investigation
Outline landfill design
Planning and disposal licence consultations etc

(b) *Site Development*
Acquisition of site
Surface water diversion/re-routing
Groundwater cut-off/separation
Lining—natural/artificial
Leachate collection/removal/treatment
Earthworks—topsoil strip/store
—bunds/embankments/roads
—cover material stockpile
Other—access roads
—screening/fencing/planting
—office/laboratory
—utilities
—wheel cleaner/weighbridge/garages etc
Gas—monitoring installations
—control
—utilization
Detailed design

(c) *Site Operation*
Operatives—wages/salaries
Plant—lease/repair/maintenance
—fuel/oil
Imported cover material
Site maintenance
Environmental—control, e.g. pests/litter
—monitoring, e.g. water/gas
Surveying
Leachate—treatment/disposal charges
Gas—venting/pumping
Rates

(d) *Site Restoration*
Capping—clay
—subsoil replacement
—topsoil replacement
—conditioning/nutrient addition
Drainage—runoff
Planting—grass
—hedges/trees
Cultivation and maintenance

(e) *Site Aftercare (for up to 15 years)*
Maintenance and cultivation
Differential settlement and treatment
Leachate—treatment/disposal
Gas—venting/utilisation
Environmental monitoring—leachate/gas
groundwater

HAZARDOUS WASTES

Costs relating to hazardous wastes handling and disposal vary widely, depending on whether the waste can be co-disposed in a landfill or whether it can be incinerated or treated in any other way, e.g. stabilization.

Depending on the type of waste and its amenability to treatment, costs can range from those shown above for landfill disposal to five hundred or even a thousand times that figure. For instance, a one-tonne load of solid-state capacitors contaminated with PCBs could cost about £1800 to incinerate, plus the handling and transport charges involved in transferring it to the incinerator.

6. LEGISLATION

INTRODUCTION

Current UK legislation with regard to solid waste exists primarily to give planning and pollution control. Additional legislation exists to monitor the movement and final disposal of hazardous wastes. Some, but not all the legislation relevant to solid-waste management, is listed in the following sections. It should be noted that for Scotland and Wales some of the legislation has different dates to those shown, and in some cases may not apply.

LEGISLATIVE BACKGROUND

Public Health Act 1848 (superseded)

Although some local environmental legislation existed before this date, the Public Health Act 1848 was the first attempt at national legislation in the UK for the control of the environmental effects of waste. It was this Act which created the term 'statutory nuisance' which refers, amongst other things, to any accumulation or deposit which is prejudicial to health or is a nuisance. The Act gave local government authorities the power to act on behalf of the public.

Public Health Act 1936 (still in force)

A series of Acts, similar to the Public Health Act 1848, culminated in the Public Health Act of 1936 which consolidated the previous legislation. In general terms this Act gave local authorities the powers to police and inspect waste arisings in their area, remove or require the removal of wastes and to prosecute offenders.

More specifically, the Act gave local authorities power to remove house and trade refuse and to require the removal of 'any accumulation of noxious matter'. It imposed a duty on local authorities to inspect their areas for 'statutory nuisances' and gave subsequent power to serve abatement notices and prosecute offenders.

Town and Country Planning Act 1947 (superseded)

The first measures to prevent damaging environmental effects of waste, rather than to clean up afterwards, were included in this Act. The Act required that any new development, including waste disposal sites and plants, should have planning permission.

Government Working Groups

As a result of public concern in the 1960s about the environmental effect of wastes, two government working groups were set up to examine the issue in more detail than formerly. The first committee, the Key Committee, was set up in 1964 to examine toxic waste; the second, the Sumner Committee, in 1967 to examine refuse disposal. The reports of these two committees influenced the drafting of Part I of the Control of Pollution Act 1974 which deals extensively with both domestic and industrial waste. A third government investigation into waste, the House of Commons Toxic Waste Inquiry, was completed in early 1989.

Town and Country Planning Act 1971 (as amended) (still in force)

This Act superseded the Town and Country Planning Act 1947,and controls the use of land for waste disposal through:

(a) The use of structure plans which set out the policies and general proposals for the development and other use of the land;

(b) Local plans which relate the policies of the structure plan to precise areas of land: and

(c) The granting or refusal of planning permission.

Deposit of Poisonous Waste Act 1972 (superseded)

An incident arising from the casual tipping of cyanide wastes in the industrial Midlands of England was the first and best known in a series of well-publicized incidents relating to the tipping of toxic wastes. This occurred prior to the drafting of Part I of the Control of Pollution Act and, whilst this drafting was in progress, interim emergency legislation was rushed through Parliament in response to public pressure for tighter controls in this area. This legislation was the Deposit of Poisonous Waste Act 1972 and, although intended to be an interim measure, was only repealed in 1981 after the implementation in 1980 of the Control of Pollution (Special Waste) Regulations 1980.

The 1972 Act was intended to 'penalize the depositing on land of poisonous, noxious or polluting wastes so as to give rise to an environmental hazard, and to make offenders liable for any resultant damage; to require the giving of notices in connection with the removal or deposit of waste, and for connected purposes'. For a waste to be considered as potentially giving rise to a hazard it has to be present in sufficient concentration or quantity to threaten death or injury to persons or animals or, alternatively, to threaten to contaminate a surface or underground water supply.

In order to monitor the movement of 'poisonous, noxious or polluting wastes' the Act established a notification procedure whereby local government and regional water authorities had to be notified both in the area from which the waste was removed and the area in which it was to be deposited. The notification procedure was applied on the exclusion principal, i.e. it applied to every waste with the exception of those listed in the Act or deposited in a manner detailed by the Act. For all wastes not listed in the Act, the notices had to specify:

(i) The premises from which it was to be removed and the land on which it was to be deposited;

(ii) The nature and chemical composition;

(iii) The quantity to be removed or deposited, together with details of the number, size and description of any containers; and

(iv) The name of the person who was to undertake the removal.

Control of Pollution Act 1974 (still in force)

The Control of Pollution Act (COPA) 1974 was intended to provide legislation for a systematic and coordinated approach to waste collection and disposal. With certain specified exemptions, the Act gave control to local (waste disposal) authorities of wastes arising from household, industrial, trade and commercial premises. These wastes are hereafter referred to as 'controlled wastes'.

Waste disposal authorities are required by the Act to draw up and regularly review a plan for the disposal of all controlled wastes. The plan should include information about:

(i) The kinds and quantities of waste which will arise in the area, or be brought into it, during the period of the plan;

(ii) What waste the authority expects to dispose of itself;

(iii) What waste others are expected to dispose of;

(iv) The methods of disposal, e.g. reclamation, incineration, landfill;

(v) The sites and equipment being provided; and

(vi) The cost.

The authorities are also required to consider what steps may reasonably be taken to reclaim and recycle waste materials.

A further major requirement of the Act is that a site should be licensed before it can be used for the treatment or disposal of controlled waste. The licence is additional to the planning permission which is required under the Town and Country Planning Act 1971, although it may only be refused if the authority is satisfied that its refusal is necessary for preventing pollution or danger to public health. All current site licences are kept on a public register, and the authority has a duty to supervise licensed activities.

Finally, following the precedent set by the Deposit of Poisonous Waste Act 1972, the 1974 Act controls discharges to water arising from waste treatment and disposal.

Control of Pollution (Special Waste) Regulations 1980 (still in force)

These special regulations for the control of toxic waste were made under Section 17 of COPA 1974 and replaced many of the provisions made in the Deposit of Poisonous Waste Act 1972. Whilst the 1972 Act operated on the 'exclusive' basis, the Section 17 Regulations operate on an 'inclusive' basis, controlling a newly-defined category of wastes called 'special wastes'. These controls are principally aimed at protecting public health rather than the environment. The environment was felt to be adequately protected by provisions in other Sections of the 1974 Act, namely by control of site licence conditions and of discharges to water.

Waste is 'special' if it is a medicinal product available only on prescription or a waste containing any of the substances listed in a schedule of the Regulations in such concentrations as to have:

(i) The ability to be likely to cause death or serious damage to tissue if a single dose of not more than 5 cm^3 was to be ingested by a child of 20 kg body weight; or

(ii) The ability to be likely to cause serious damage to human tissue by inhalation, skin contact or eye contact on exposure to the substance for 15 minutes or less; or

(iii) A flash point of 21°C or less.

The Regulations provide for a 'cradle to grave' control system for the disposal of special wastes. A consignment system has been set up, and waste producers are required to inform receiving waste disposal authorities of their intention to dispose of special wastes. Compliance with the consignment system requires a record to be kept of the dispatch, conveyance and disposal of the waste, with a permanent record being retained at the place of ultimate disposal.

Refuse Disposal (Amenity) Act 1978 (still in force)

This Act places a duty on local authorities to provide sites to which residents of that area may bring bulky household refuse free of charge.

Dumping at Sea Act 1974 (superseded)

Former voluntary schemes covering nearly all industrial waste and sewage sludges were formally implemented by the Dumping at Sea Act 1974. This Act made it an offence to dump, or to load for the purpose of dumping, any material in UK waters or from a UK ship, vehicle, hovercraft or marine structure without a licence from the relevant licensing authority. The licensing authority is required by the Act to 'have regard to the need to protect the marine environment and the living resources which it supports from any adverse consequence of dumping'. The term 'Dumping at Sea' includes incineration at sea.

Food and Environment Act 1985 (still in force)

Part II of the Food and Environment Act 1985 supersedes the Dumping at Sea Act 1974, but provides similar legislative control.

EC Legislation

There are a number of EC Directives on Wastes Management. These are listed in Table VIII, and have generally been implemented in the UK under COPA 1974, and more specifically under the Control of Pollution (Special Waste) Regulations 1980.

COPA 1974 lays down particular requirements for the protection of ground and surface waters. Groundwaters are protected from the polluting effects of waste disposal by *EEC Directive 80/68* which was complied with mainly in the UK under COPA 1974. This Directive has appended List I and List II of families and groups of dangerous substances, those on List I generally being more dangerous than those on List II. The Directive requires the prevention of List I substances from, and a limit to the quantity of List II substances, entering ground water. All direct discharges of List I substances are to be prohibited, although if after investigation the groundwater is found to be unsuitable for other uses, such discharges may be authorized.

EFFECTS ON UK PRACTICE

Current UK practice

Landfill practice in the UK has altered considerably, particularly as a result of the Deposit of Poisonous Waste Act 1972 and COPA 1974 (described above). This has resulted in restrictions in the type of wastes that can be placed in landfill sites, and has allowed local authorities a means (through planning and site licence consents) of further control. Thus the type of wastes and manner of placing, together with environmental aspects such as gas and leachate control, and eventual site restoration can all be stipulated by the local authority.

The restrictions which can now be placed on a landfill site operation mean that, prior to applying for planning permission or a site licence, it is incumbent upon the proposed

Table VIII. EC legislation on wastes management

No	*Description*	*Formal compliance in UK*
75/442/EEC	Directive on waste	July 1977 under COPA; actually implemented July 1978
78/319/EEC	Directive on toxic and dangerous waste	22nd March 1980 under COPA and 1981 under Section 17 (Special Waste Regs)
84/361/EEC	Directive on the supervision and control within the European Community of the transfrontier shipment of hazardous waste	November 1988
85/469/EEC	Adaptation (Commission Directive)	
86/279/EEC	Amendment (Council Directive)	
87/112/EEC	Adaptation (Commission Directive)	
76/403/EEC	Directive on the disposal of polychlorinated biphenyls and polychlorinated terphenyls	1981 under Section 17 Regs
86/278/EEC	Directive on the protection of the environment and in particular of the soil, when sewage sludge is used in agriculture	19th June 1989

operator of the site to carry out a full site investigation and engineering design. The site investigation should particularly identify any potential geological pathways for pollutants such as fissures or permeable and porous strata. The engineering design should take account of the potential of waste to produce pollutants, and should aim to limit their production and control their eventual discharge into the environment.

All leachate discharges are controlled: to groundwater under the *EEC Directive 80/68* and to surface and marine waters under COPA II 1974. There is thus a control on the type and quantity of pollutants which can enter any watercourse, and the opportunity for local authorities to prohibit any such discharge if it is deemed appropriate. This has resulted in the increasing use of contained (i.e. naturally or artificially impermeably lined) sites, limiting discharges to ground water, and of the practice of collecting and treating leachate — limiting discharges to surface waters.

Fly tipping, i.e the unwanted, indiscriminate dumping of wastes, is a practice which has been a problem in the UK for centuries. Each successive Act aimed at curbing the damaging environmental effects of waste management has sought to control fly tipping. Whilst it is still difficult to apprehend the casual fly tipper, a curb has been placed on the fly tipping of particularly toxic substances. This curb comes in the form of the *Control of Pollution (Special Waste) Regulations 1980* where the consignment procedure should enable a load of toxic waste to be traced back to the transporter if not properly tipped.

With a growing shortage of landfill sites in locations which are convenient for waste disposal, increasing use is being made in the UK of technologies such as incineration and pulverization/shredding and in the production of refuse-derived fuels. A wide variety of further legislative controls exists for these operations, but will not be detailed here except to note that all these operations require planning permission.

Her Majesty's Inspectorate of Pollution

In 1986 the UK Government announced its decision to establish a new unified Pollution Inspectorate for England and Wales. The new inspectorate, Her Majesty's Inspectorate of Pollution (HMIP), was required to develop a more coherent approach to the control of industrial emissions to air, water or land and to provide advice which will help the other pollution control authorities to carry out their statutory responsibilities. This Inspectorate comprises the former Alkali Inspectorate (responsible for atmospheric emissions) the Hazardous Waste Inspectorate (formed in 1981 to examine and make recommendations for the management of hazardous waste) and a newly-formed Water Directorate (responsible for discharges to water). It should be noted that, whereas the Hazardous Waste Inspectorate concerned itself only with hazardous wastes, Her Majesty's Inspectorate of Pollution is required to examine the management of all controlled wastes.

GLOSSARY

There are many categories of waste, although for the purposes of this book they are broken down into three broad bands: (i) non-hazardous, (ii) hazardous and (iii) radioactive. Agricultural wastes and sewage sludge are included here only as regards their ultimate disposal. The management of sewage sludge has been otherwise dealt with in an IWEM booklet entitled *An Introduction to Sewage Treatment*.

Within the three broad bands defined above there is a further breakdown of types of waste, for instance for solid waste these include domestic, commercial and industrial. This further breakdown of the categories is given below, with each category briefly defined.

NON-HAZARDOUS WASTES

Household Waste

This is waste which is produced by private residential homes, by educational and nursing (including hospitals—but not clinical wastes) establishments. It comprises all waste produced as part of day-to-day living. A typical analysis is given in on page 14.

Civic Amenity

Civic-amenity waste is the bulky waste taken by householders to civic amenity collection sites.

Commercial Waste

Waste from premises used wholly or mainly for the purpose of a trade or business or for the purpose of sport, recreation or entertainment.

Industrial Waste

Waste from a factory or industrial premises. This category excludes mine and quarry wastes but includes construction waste, dredging spoil and sewage sludge (see below).

Trade Effluent

Waste substances—usually liquid in nature—arising from a factory or industrial process. This may be hazardous or non-hazardous. (In the context of this book only hazardous trade effluents are considered and, of these, only those which are not disposed of, without treatment, to sewer.) In recent years there has been a general policy to recycle and recover as many as possible of the by-products of an industrial process. These trade effluents are therefore more similar to a sludge than a liquid, and are fairly toxic in nature.

Controlled Waste

This is a term defined by the Control of Pollution Act 1974 (see Section 6) and includes household, industrial and commercial waste.

Difficult Waste

A term used by local authorities for wastes which require special care in their handling, treatment or disposal. This is a wider category than hazardous waste and some, but not all, difficult wastes may be hazardous.

Putrescible Wastes

This is a term applied to wastes which can very easily be broken down by biological action. Typical wastes which are putrescible are food and vegetable wastes in domestic waste, sewage sludge, agricultural manure wastes, wastes from food-processing factories, tanneries, glue factories. Wastes which have a high cellulose content such as rice and wheat husks, paper and wood, although they are biodegradable, are not generally considered to be putrescible.

Inert Wastes

Inert wastes are those which are biologically and chemically stable. Depending on the precise definition used by the waste disposal authority, this category is generally taken to include building wastes and sub-soils. However, contaminants can be leached out (chemical action) of the building wastes, and the organic matter in soil and sub-soil can degrade (biological action).

Sewage Sludge

Sewage sludge results from the treatment of sewage. The characteristics of the sludge will depend on the method of treatment employed. Sludge from primary settling tanks has less water than other sludges (92–97 per cent) but may contain high concentrations of metals and bacteria. Sludge from humus tanks or clarifiers is more flocculent, but may also be more filamentous in nature and difficult to dewater than primary sludge. It may have similar or lower heavy metal and bacteria concentrations. For this reason it is often co-settled with primary sludge. Any sludge treated in an anaerobic digester has a reduced organic content and is more pleasant to handle. It is difficult to dewater mechanically but dries easily on beds. Sludge which has been treated with coagulants has a significantly lower water content than other sludges.

Agricultural Wastes

This term applies to manures produced by farm animals. It may also apply to arable wastes such as straw if these are disposed of other than by burning or composting on site. Waste pesticides and fertilizers are categorized as agricultural waste whilst they remain (or are disposed of) on the farm but, because of their chemical nature, are generally classified as Special Wastes if the farmer requires to dispose of them in a responsible manner off his land.

HAZARDOUS WASTES

General Definition

All wastes which have a harmful effect on the environment and on human health. This category includes special wastes but is also extended to other wastes not so defined. The various categories which are considered in this booklet are defined below and are limited to those for which a specific treatment or disposal route is required.

Special Wastes

These are defined under the Control of Pollution Act 1974 as waste which "is or may be so dangerous or difficult to dispose of that special provision . . . is required for [its] disposal".

It is further defined as having a flashpoint of 21°C or less, of being capable of causing serious damage or death by ingestion (5 cm^3 consumed by a child of 20 kg) or of damage to skin tissue by exposure for 15 minutes or less.

PCBs

Polychlorinated biphenyls (PCBs) were originally added to transformer oils to improve electrical conductivity. They have since been found to persist in the environment and to produce extremely toxic substances—dioxins and dibenzofurans—on partial combustion. A ban was placed on their sale or use in 1986. The waste may be in the form of liquids, solids or sludges.

Acid Tars

Acid tars are produced from the acid washing of benzene, toluene and xylene fractions distilled from crude coal carbonization of benzole, from the production of lubricating oils and from refining waste oils. Principally a liquid waste, as produced, in some a sludge settles out on storage making handling difficult. They generally comprise 65–90 per cent sulphuric acid, up to 15 per cent organic compounds and up to 20 per cent water.

Metal-Finishing Wastes

These wastes are in the form of both liquids and sludges, and comprise acid pickle liquors and metal hydroxide sludges. They are inorganic.

Halogenated Hydrocarbon Solvents

These wastes may be in the form of a solid residue, a liquid oily waste and a sludge.

Cyanide Wastes

Cyanide is toxic in certain forms, and less toxic in others. It was the discovery of indiscriminantly-dumped cyanide wastes (see Section 6) that led to the Deposit of Poisonous Waste Act 1972. Cyanide wastes result from coal carbonization, metal plating and hardening, and are in the form of solid residues and liquors.

Clinical Wastes

These are principally wastes from hospitals and nursing homes, and include bandages, dressings, syringes and various organic substances. They are not officially classed as hazardous wastes but require special handling and disposal, generally by incineration. They are specifically covered under the *Collection and Disposal of Waste Regulations 1988.*

RADIOACTIVE WASTES

Radioactive Waste

Waste contaminated by reason of its radioactivity arising from a variety of sources including hospitals, nuclear power plants and defence installations. For disposal purposes, radioactive wastes are defined according to their sources. Any waste with a radioactivity of less than 10^{-2} microcuries per tonne is considered to be non-active.

Low-Level Radioactive Waste

There is no precise definition but NIREX, based at Harwell, propose an upper limit of approximately 200–500 microcuries per tonne for beta and gamma radiation and of 100 microcuries per tonne for alpha radiation. These wastes are usually materials such as used gloves and syringes.

Intermediate-Level Radioactive Waste

Again, there is no precise definition, and activity ranges upwards from the upper limit of the low-level material. These wastes are generally 1000 times more active than low-level wastes.

High-Level Radioactive Waste

Generally radioactive waste which generates heat is classed as high-level waste. These wastes normally only arise when spent nuclear fuel is dissolved in acid to separate the waste (fission products) from the re-usable uranium. The waste generates a lot of heat and cannot be acceptably disposed of by containment.

ABBREVIATIONS USED IN TEXT

CIPFA	Chartered Institute of Public Finance and Accountancy
COPA	Control of Pollution Act, 1974
DAFS	Department of Agriculture and Fisheries for Scotland
DoE	Department of the Environment
HWI	Hazardous Waste Inspectorate
ISO	International Standards Organization
MAFF	Ministry of Agriculture, Fisheries and Food
NIREX	Nuclear Industry Radioactive Waste Executive
NRA	National River Authority
PCB	Polychlorinated biphenyl
RDF	Refuse-derived fuel
RPB	River Purification Board (Scotland)
WA	Water Authority
WDA	Waste Disposal Authority

Bibliography and Some Further Sources of Information

Publications by the DoE, available from HMSO:

1. Waste Management Papers

Waste Management Paper No.	Date of Publication	Title
1	*1976*	*Reclamation, Treatment and Disposal of Wastes—An Evaluation of Options*
2	*1976*	*Waste Disposal Surveys*
3	*1976*	*Guideline for the Preparation of a Waste Disposal Plan*
4	*1st issued 1976* *re-issued 1988*	*The Licensing of Waste Disposal Sites*
5	*1976*	*The Relationship between Waste Disposal Authorities and Private Industry*
6	*1976*	*Polychlorinated Biphenyl (PCB) Wastes—A Technical Memorandum on Reclamation, Treatment and Disposal including a Code of Practice*
7	*1976*	*Mineral Oil Wastes—A Technical Memorandum on Arisings, Treatment and Disposal including a Code of Practice*
8	*1976*	*Heat-treatment Cyanide Wastes—A Technical Memorandum on Arisings, Treatment and Disposal including a Code of Practice*
9	*1976*	*Halogenated Hydrocarbon Solvent Wastes from Cleaning Processes—A Technical Memorandum on Reclamation and Disposal including a Code of Practice*
10	*1976*	*Local Authority Waste Disposal Statistics 1974–75*
11	*1976*	*Metal Finishing Wastes—A Technical Memorandum on Arisings, Treatment and Disposal including a Code of Practice*
12	*1977*	*Mercury Bearing Wastes—A Technical Memorandum on Storage, Handling Treatment and Recovery of Mercury including a Code of Practice*
13	*1977*	*Tarry and Distillation Wastes and Other Chemical Based Residues—A Technical Memorandum on Arisings, Treatment and Disposal including a Code of Practice*
14	*1977*	*Solvent Wastes (excluding Halogenated Hydrocarbons)—A Technical Memorandum on Reclamation and Disposal including a Code of Practice*

15	*1978*	*Halogenated Organic Wastes—A Technical Memorandum on Arisings, Treatment and Disposal including a Code of Practice*
16	*1980*	*Wood Preserving Wastes—A Technical Memorandum on Arisings, Treatment and Disposal including a Code of Practice*
17	*1978*	*Wastes from Tanning Leather Dressing and Fellmongering—A Technical Memorandum and Recovery Treatment and Disposal including a Code of Practice*
18	*1979*	*Asbestos Waste—A Technical Memorandum on Arisings and Disposal including a Code of Practice*
19	*1978*	*Wastes from the Manufacture of Pharmaceutical, Toiletries and Cosmetics—A Technical Memorandum on Arisings and Disposal including a Code of Practice*
20	*1980*	*Arsenic Bearing Wastes—A Technical Memorandum on Recovery, Treatment and Disposal including a Code of Practice*
21	*1980*	*Pesticide Wastes—A Technical Memorandum on Recovery, Treatment and Disposal including a Code of Practice*
22	*1979*	*Local Authority Waste Disposal Statistics 1974–75 to 1977–78*
23	*1981*	*Special Wastes—A Technical Memorandum Providing Guidance on their Definition*
24	*1984*	*Cadmium Bearing Wastes—A Technical Memorandum on Arisings, Treatment and Disposal including a Code of Practice*
25	*1983*	*Clinical Waste—A Technical Memorandum on Arisings, Treatment and Disposal including a Code of Practice*
26	*1986*	*Landfilling Wastes—A Technical Memorandum for the Disposal of Wastes on Landfill Sites*
27	*1989*	*The Control of Landfill Gas—A Technical Memorandum on the Monitoring and Control of Landfill Gas*

Reports of the Hazardous Waste Inspectorate

First Report	*June 1985*	*Hazardous Waste Management: An Overview*
Second Report	*June 1986*	*Hazardous Waste Management: Ramshackle & Antediluvian*
Third Report	*June 1988*	

Reports of Her Majesty's Inspectorate of Pollution

First Report	*March 1989*	*HMIP First Annual Report 1987–88*

Reports of the Royal Commission on Environmental Pollution

Eleventh Report	*December 1985*	*Managing Waste: the Duty of Care. Cmnd. 9675*
Twelfth Report	*February 1988*	*Best Practicable Environmental Option. Cmnd. 310*

Government Committees

July 1981	*House of Lords Select Committee on Science and Technology: Session 1980–81, First Report: Hazardous Waste Disposal*
November 1984	*Report of the Independent Review of Disposal of Radioactive Waste in the North East Atlantic*
February 1989	*House of Commons Session 1988–89 Environment Committee Second Report: Toxic Waste*

Consultation Papers

July 1988	*Integrated Pollution Control*
January 1989	*The Role and Functions of Waste Disposal Authorities*

2. Other Useful Publications

HAIGH, N.	*EEC Environmental Policy & Britain.* 2nd Edition 1987. Longman Goup UK Ltd
HOLMES, J. E. (Ed)	*Practical Waste Management.* 1983. John Wiley & Sons
WILSON, D. C.	*Waste Management: Planning, Evaluation, Technologies.* 1981. Oxford University Press
PORTEOUS, A. (Ed)	*Hazardous Waste: Management Handbook.* 1985. Butterworths
INSTITUTION OF WATER AND ENVIRONMENTAL MANAGEMENT	*An Introduction to Sewage Treatment**. 1987.
Ibid	*Sewage Sludge I: Production, Preliminary Treatment and Digestion**. 1979.
Ibid	*Sewage Sludge II: Conditioning, Dewatering and Thermal Drying**. 1981.
Ibid	*Sewage Sludge III: Utilization and Disposal**. 1978.

* *Copies of these publications are obtainable from the Institution's Headquarters.*

INDEX

Printed in Great Britain by Headley Brothers Ltd The Invicta Press Ashford Kent and London